Boston Studies in the Philosophy and History of Science

Volume 311

Alisa Bokulich
Department of Philosophy
Boston University
Boston
Massachusetts
USA

Robert S. Cohen
Boston University
Watertown
Massachusetts
USA

Jürgen Renn
Max Planck Institute for the History of Science
Berlin
Germany

Kostas Gavroglu
University of Athens
Athens
Greece

The series *Boston Studies in the Philosophy and History of Science* was conceived in the broadest framework of interdisciplinary and international concerns. Natural scientists, mathematicians, social scientists and philosophers have contributed to the series, as have historians and sociologists of science, linguists, psychologists, physicians, and literary critics.

The series has been able to include works by authors from many other countries around the world.

The editors believe that the history and philosophy of science should itself be scientific, self-consciously critical, humane as well as rational, sceptical and un-dogmatic while also receptive to discussion of first principles. One of the aims of Boston Studies, therefore, is to develop collaboration among scientists, historians and philosophers.

Boston Studies in the Philosophy and History of Science looks into and reflects on interactions between epistemological and historical dimensions in an effort to understand the scientific enterprise from every viewpoint

More information about this series at http://www.springer.com/series/5710

William J. Devlin • Alisa Bokulich
Editors

Kuhn's Structure of Scientific Revolutions—50 Years On

 Springer

Editors
William J. Devlin
Department of Philosophy
Bridgewater State University
Bridgewater
Massachusetts
USA

Alisa Bokulich
Department of Philosophy
Boston University
Boston
Massachusetts
USA

ISSN 0068-0346   ISSN 2214-7942 (electronic)
Boston Studies in the Philosophy and History of Science
ISBN 978-3-319-37720-9   ISBN 978-3-319-13383-6 (eBook)
DOI 10.1007/978-3-319-13383-6

Printed on acid-free paper

Springer is part of Springer Science+Business Media (www.springer.com)

For Robert S. Cohen Founder of this Boston Studies book series Co-founder of the BU Center for Philosophy and History of Science and the Boston Colloquium for Philosophy of Science. In deep gratitude for your lifelong contributions to this field.

Acknowledgments

It has been an honor to develop this volume as both a contribution to the Boston Studies in the Philosophy and History of Science series and a commemoration of Thomas Kuhn's work. We are grateful to Springer Press for giving us this opportunity to help celebrate and further explore *The Structure of Scientific Revolutions*. We are also grateful to Lucy Fleet, our editor at Springer Press, for her assistance throughout the process. We are deeply thankful to our twelve contributors, whose contributions of original essays help to provide novel insights into the impact *Structure* has had, and continues to have, in the fields of philosophy, history, science, and sociology.

In addition, Bill would like to thank his colleagues in the Department of Philosophy at Bridgewater State University for their academic support and philosophical discussions about Kuhn throughout the development of this volume. He would also like to thank his family for their continued support: his parents, Robert and Margaret, as well as Rob, Katie, RJ, James, and Janice Wilson. He would lastly like to thank Richard and Adri Howey for their friendship and academic counseling.

Contents

1   **Introduction** 1
William J. Devlin and Alisa Bokulich

2   **Kuhn's *Structure*: A Moment in Modern Naturalism** 11
Steven Shapin

3   **Kuhn and the Historiography of Science** 23
Alexander Bird

4   **From Troubled Marriage to Uneasy Colocation: Thomas Kuhn,
Epistemological Revolutions, Romantic Narratives, and History
and Philosophy of Science** 39
Alan Richardson

5   **Reconsidering the Carnap-Kuhn Connection** 51
Jonathan Y. Tsou

6   **The Rationality of Science in Relation to its History** 71
Sherrilyn Roush

7   **What do Scientists and Engineers Do All Day? On the Structure of
Scientific Normalcy** 91
Cyrus C. M. Mody

8   **From Theory Choice to Theory Search: The Essential Tension
Between Exploration and Exploitation in Science** 105
Rogier De Langhe and Peter Rubbens

9   **The Evolving Notion and Role of Kuhn's Incommensurability Thesis** 115
James A. Marcum

10   **Walking the Line: Kuhn Between Realism and Relativism** 135
Michela Massimi

11 An Analysis of Truth in Kuhn's Philosophical Enterprise 153
William J. Devlin

12 Kuhn's Social Epistemology and the Sociology of Science 167
K. Brad Wray

13 Kuhn's Development Before and After *Structure* 185
Paul Hoyningen-Huene

Index 197

Contributors

Alexander Bird Department of Philosophy, University of Bristol, Bristol, UK

Alisa Bokulich Center for Philosophy and History of Science, Department of Philosophy, Boston University, Boston, MA, USA

Rogier De Langhe Tillburg Center for Logic and Philosophy of Science (TiLPS), Tilburg University, Tilburg, The Netherlands

William J. Devlin Department of Philosophy, Bridgewater State University, Bridgewater, MA, USA

Paul Hoyningen-Huene Institute of Philosophy and Center for Philosophy and Ethics of Science, Leibniz Universität, Hannover, Germany

James A. Marcum Department of Philosophy, Baylor University, Waco, USA

Michela Massimi School of Philosophy, Psychology and Language Sciences, University of Edinburgh, Edinburgh, Scotland, UK

Cyrus C. M. Mody Department of History, Rice University, Houston, TX, USA

Alan Richardson Department of Philosophy, The University of British Columbia, Vancouver, BC, Canada

Sherrilyn Roush Department of Philosophy, King's College London, London, UK

Peter Rubbens Department of Physics and Astronomy, Ghent University, Ghent, Belgium

Steven Shapin Department of the History of Science, Harvard University, Cambridge, MA, USA

Jonathan Y. Tsou Department of Philosophy and Religious Studies, Iowa State University, Ames, IA, USA

K. Brad Wray Department of Philosophy, State University of New York, Oswego, NY, USA

Chapter 1
Introduction

William J. Devlin and Alisa Bokulich

In Spring of 2012, the Center for Philosophy and History of Science at Boston University hosted a session of the Boston Colloquium celebrating the 50th anniversary of Thomas Kuhn's *Structure of Scientific Revolutions* (henceforth, *Structure*).This colloquium brought together many of the world's leading scholars on Kuhn and we were honored to have the Kuhn family present as well to share their own personal stories and perspectives about Thomas Kuhn and his work. It was not the first time the Center had marked an anniversary of this book: In 1982 Kuhn himself had come to speak in the Boston Colloquium about his *Structure* 20 years on.What is remarkable about this book, which has sold well over a million copies, is that it continues to shape the history and philosophy of science and popular thinking as much today, 50 years on, as it did 20 years on, though perhaps not in exactly the same way. Exploring just what the legacy of Kuhn's *Structure* is 50 years on is the subject of this volume. While the inspiration for this volume came from this colloquium, several additional scholars representing diverse perspectives within the history and philosophy of science have contributed papers.

As Kuhn himself explained, the central ideas of *Structure* were motivated by his time as a graduate student in the summer of 1947 when he attempted to read, and understand, Aristotle's *Physics* through the perspective of Newtonian mechanics. Realizing Aristotle knew nearly nothing about mechanics, Kuhn eventually came to the view that Aristotelian science was not simply a system of "egregious mistakes",

W. J. Devlin (✉) · A. Bokulich
Department of Philosophy, Bridgewater State University,
45 School Street, Bridgewater, MA 02325, USA
e-mail: wdevlin@bridgew.edu

A. Bokulich
Center for Philosophy and History of Science, Department of Philosophy,
Boston University, 745 Commonwealth Avenue, Boston, MA 02128, USA
e-mail: abokulic@bu.edu

© Springer International Publishing Switzerland 2015
W. J. Devlin, A. Bokulich (eds.), *Kuhn's Structure of Scientific Revolutions—50 Years On*,
Boston Studies in the Philosophy and History of Science 311,
DOI 10.1007/978-3-319-13383-6_1

but that it belonged to a self-contained and successful scientific tradition. This idea
gave rise to Kuhn's famous concept of *scientific revolutions* (Kuhn, [1987] 2000).[1]

There are at least three central themes in *Structure*. One is that science operates
within *paradigms*, which Kuhn describes as consisting of unprecedented scientific
achievements that are sufficiently open-ended and successful to attract a group of
adherents to focus on resolving the remaining problems (Kuhn [1962] 1996, §V).
These achievements are "accepted examples of actual scientific practice—examples
which include law, theory, application, and instrumentation together—[and] provide
models from which spring particular coherent traditions of scientific research" (Kuhn
1970, p. 10). Due to equivocal uses of the term 'paradigm' in *Structure*,[2] Kuhn later
provides a 'global' and a 'local' definition of paradigm. He defines paradigm globally as a *disciplinary matrix*, insofar as it is "embracing all the shared commitments
of a scientific group" (Kuhn 1974, p. 460), which includes (1) symbolic generalizations: formalizable components of the paradigm, including laws and definitions
(e.g., f=ma); (2) metaphysical beliefs (such as 'heat is the kinetic energy of the
constituent parts of bodies'); (3) values (such as accuracy, consistency, breadth of
scope, simplicity, and fruitfulness); and (4) shared exemplars (Kuhn [1962] 1996, pp.
182–186). This last element, *shared exemplars*, is the second, more local notion of
paradigm, and the one that Kuhn thought was more fundamental. A shared exemplar
is understood as a set of "concrete problem solutions, accepted by the [disciplinary]
group as, in a quite usual sense, paradigmatic" (Kuhn [1962] 1996, pp. 186–187).
They are models of how members of a disciplinary matrix should conduct scientific
investigations, including theories and their application, the method of observation
(i.e., what they are looking for in their experiments), and which instruments to use
and how to apply them.

Another central theme of *Structure* is the historical pattern of theory change in the
development of science.[3] Kuhn explains that paradigm, as shared exemplar, is crucial
to the normal activities of science, or what he calls *normal science*:[4] that is, "research
firmly based upon one or more past scientific achievements, achievements that some
particular scientific community acknowledges for a time as supplying the foundation
for its further practice" (Kuhn [1962] 1996, p. 10). Normal science perpetuates the
acceptance of the paradigm as a disciplinary matrix until the scientists practicing
normal science confront an *anomaly* (§VI), that is, a piece of the puzzle that doesn't
seem to fit into the framework determined by the shared exemplars. If unresolved, an
anomaly brings forth a *crisis* (§VII-VIII), which leads to the possibility of shifting
the problem-solving techniques of normal science and perhaps even the destruction

---

[1] For a further discussion of the historical background to Structure, see Shapin, Chap. 2, in this
volume.

[2] See Margaret Masterman (1970) and Dudley Shapere (1964) for the problem of Kuhn's multiple
uses of the term 'paradigm'. Kuhn's response and clarification of 'paradigm', which avoids problems
of ambiguity, is found in the 1969 Postscript to Structure, along with his (1974).

[3] For a further discussion of Kuhn and the notion of scientific change, see Bird, Chap. 3, in this
volume.

[4] For an examination of the nature of normal science today, see Mody, Chap. 7, in this volume.

of the global paradigm. In the latter case, we have what Kuhn calls a *revolution* (§IX) in science, that is, "non-cumulative developmental episodes" in which the anomalies lead to a crisis and result in the replacement of "an older paradigm in whole or in part by an incompatible new one" (Kuhn [1962] 1996, p. 92). Mature science, then, develops through a regularly patterned cycle of phases: paradigm—normal science—anomaly—crisis—revolution—new paradigm.[5]

While the resolution of a revolution entails a paradigm-shift, the comparison between an old and new paradigm leads to the third and, as James Conant and John Haugeland describe, the "most widely criticized" (Kuhn 2000, p. 4) central theme of *Structure*: namely, the concept of *incommensurability* (§§XI-XII). As Kuhn explains, "[t]he normal-scientific tradition that emerges from a scientific revolution is not only incompatible but often actually incommensurable with that which has gone before" (Kuhn [1962] 1996, p. 103). Kuhn understands his incommensurability thesis to involve three key elements: first, proponents of competing paradigms will typically not agree on the list of problems a paradigm must solve, nor on the appropriate standards for their resolution. Second, although many of the same terms will appear in both the old and new paradigms, they will not often have the same meaning (such as 'space' for Newton versus 'space' for Einstein). Kuhn explains "the third and most fundamental aspect of the incommensurability [is that] proponents of competing paradigms practice their trades in different worlds: (Kuhn [1962] 1996, p. 150). Kuhn is quick to note that this does not mean scientists can see anything they please, but there can be a sense in which "the data themselves" have changed with a change of paradigm (Kuhn [1962] 1996, p. 135). Following N.R. Hanson (1958), Kuhn notes that the perceptual experiences that an observer has are not uniquely determined by the objects of perception alone; rather, our background theories influence how we understand our visual experiences. So, for instance, when looking at swinging stones, scientists of the Aristotelian paradigm see constrained fall, while scientists of the Galilean paradigm see pendulum motion. Hence, he concludes that it is as if scientists of different paradigms are living in different and incommensurable worlds—or as he puts it more measuredly at other times, different worlds that are "here and there incommensurable" (Kuhn [1962] 1996, p. 112).[6]

Kuhn recognized that, when taken together, these three themes of paradigms, scientific change, and incommensurability, give rise to new picture of the enterprise of science. Kuhn closes *Structure* (§XII) by rejecting the idea that science develops by accumulating truths about nature and he concludes that these central ideas lead to a rejection of scientific realism.[7] Although Kuhn is willing to grant an "instrumental"

---

[5] For a discussion of these universalist and normative elements in Kuhn's historical project, see Bird, Chap. 3, in this volume and specifically in connection with questions about the relationship between history of science and philosophy of science, see Richardson, Chap. 4, in this volume.

[6] For an analysis of Kuhn and Carnap concerning the notion of incommensurability, see Tsou, Chap. 5, in this volume. For a further discussion of Kuhn's different notions of incommensurability, see Marcum, Chap. 9, in this volume.

[7] See Massimi, Chap. 10, in this volume for a further discussion of Kuhn and scientific realism, along with Kuhn's relativism.

progress in science across revolutions, that is, that "the list of problems solved by science and the precision of individual problem-solutions will grow and grow [we must] relinquish the notion, explicit or implicit, that changes of paradigm carry scientists closer and closer to the truth" (Kuhn [1962] 1996, p. 170).[8]

The central ideas of *Structure* have undeniably had a profound effect on studies in the history and philosophy of science over the past 50 years. However, *Structure* has certainly not been without its critics. Not long after its publication, it was beset with philosophical criticism: for example, Dudley Shapere, Israel Scheffler, and Karl Popper each maintained that Kuhn's account of scientific change entails that theory-choice is subjective and irrational.[9] Furthermore, Shapere and Popper argued that by rejecting the claim that science achieves truth, and instead endorsing the claim that science operates under the notions of paradigms, revolutions, and incommensurability, Kuhn makes science relative. Although Kuhn categorically denies these ascriptions of irrationality and relativism in his 1969 Postscript to *Structure* (which became part of the second edition in 1970, and all subsequent editions), responding to these and other objections occupied Kuhn for the remainder of his life.[10] As we shall see, it has occupied the careers of many other scholars in the 50 years since *Structure* as well.

In the second chapter, Steven Shapin examines *Structure* as an argument about the nature of science. Here, Shapin focuses on understanding *Structure* by exploring the specific historical circumstances from which it originated. In particular, Shapin provides an account of the *conditions of possibility* for the naturalistic sentiments of the book—sentiments which help give rise to the central components of the work, such as paradigms, revolutions, normal science, truth, reason, reality, and progress. In his description of the historicity of these sentiments, Shapin argues that the conditions of possibility of these sentiments concern changes to the political and economic circumstances of science in the twentieth century. Shapin turns to three texts, each published a year apart from *Structure*, which help to show that science was becoming institutionally and culturally normalized: (1) President Eisenhower's Farewell Address; (2) Alvin Weinberg's paper, "Impact of Large-Scale Science on the United States"; and (3) Derek de Sollar Price's *Little Science,Big Science*. Shapin concludes that this institutionalization and cultural normalization of science served as the conditions that made Kuhn's naturalistic approach towards science in *Structure* possible.

---

[8] While Kuhn believes that the list of problems solved by science will grow in number, he notes that after a revolution it need not be the same problems that are solved; for example, while Descartes' celestial mechanics solved the problem of answering why all the planets orbit the sun in the same direction, in the transition to the Newtonian paradigm a solution to this problem was lost (such examples have come to be known as "Kuhn losses").

[9] For the charges of irrationality and subjectivity, see Shapere (1966) and Scheffler (1967).Likewise, see also Shapere's and Popper's essays contained in Lakatos and Musgrave (1970). For an analysis of the relation between Kuhn and the rationality of science, see Roush, Chap. 6, in this volume.

[10] It seems Kuhn had been planning to address at least some of these issues in his next book, The Plurality of Worlds: An Evolutionary Theory of Scientific Development, of which only an unfinished manuscript exists. For a further discussion of this manuscript, see Hoyningen-Huene, Chap. 13, of this volume.

In Chapter Three, Alexander Bird shows that Kuhn's thinking about the nature of the history of science can be illuminated by examining two distinct strands of historicism found in the work of Hegel. First, Bird maintains that Kuhn's historical approach to science—i.e., Kuhn's view that, methodologically, we examine the history of the scientific tradition and the historical context of the development of scientific ideas to understand scientific change—exemplifies Hegel's *conservative* strand of historicism. Second, Bird shows that, given Kuhn's identification of the cyclical patterns of normal and revolutionary science, Kuhn exemplifies Hegel's *determinist* strand of historicism. Furthermore, Bird holds that, together, these two strands of historicism in Kuhn's thinking complement one another. On the one hand, the determinist strand provides us with Kuhn's belief of the cyclical patterns of scientific change. On the other hand, the conservative strand explains those patterns. From this he argues that Kuhn should be understood as adopting *internalism* towards scientific change. Finally, given Kuhn's internalism, Bird shows why Kuhn rejected certain forms of social constructivism.

In Chapter Four, Alan Richardson examines how Kuhn conceived of the relationship between the history of science and the philosophy of science, and draws out its implications for our understanding of this relationship today. He points out that Kuhn was quite conscious of the different goals and interests in history and philosophy as he developed *Structure*. This suggests that the "role for history" in *Structure* is more subtle than we normally think. *Structure*, according to Richardson, should not be read primarily as a philosophical treatise that presents the shortcomings of logical empiricism and offers a new view that resolves such problems. Instead, we should understand that it begins with historiographic questions and thereby suggests that logical empiricism gives the wrong advice about what questions are important to the historian of science. Likewise, Kuhn recognized that, by holding to the view that theories are simply sets of laws, philosophers were not able to properly reconstruct the process of theory development or evaluation. In thinking about the nature of philosophy and history, however, Richardson argues that we should eschew stereotypical dichotomies, such as universal versus particular, timeless versus timebound, or normative versus descriptive.Instead, Richardson concludes that we should use Kuhn's emphasis on the practices of history and philosophy to help us think about how history of science and philosophy of science can best work together today, by better articulating their own explanatory projects.

In the fifth chapter, Jonathan Tsou examines recent scholarship on the relationship between Kuhn and Rudolf Carnap, which argues that, based on their similar views concerning incommensurability, theory-choice, and scientific revolutions, these two philosophers of science are not as dissimilar to one another as it has commonly been supposed. Tsou argues that this "revisionist view" of the relationship between Carnap and Kuhn is misguided. Although there are similarities between the two philosophical systems, Tsou maintains that these similarities are superficial. When we consider the broader philosophical projects of both thinkers, we find that there is a stark and fundamental difference in motivation between their systems: while Carnap's system is motivated by the intention to clarify the distinction between meaningful and meaningless questions, Kuhn's system is motivated to offer a naturalistic account of the history of science.This general difference further reveals specific differences between Carnap's and Kuhn's conceptions of incommensurability, theory-choice, and scientific revolutions. Tsou concludes that Kuhn's philosophical views do indeed mark a revolutionary departure from Carnap's.

In Chapter Six, Sherrilyn Roush explores Kuhn's view of the history of science in relation to questions about the rationality of science. Specifically, she examines two challenges to the rationality of science. The first challenge is the traditional criticism that Kuhn's notions of paradigm and paradigm shifts in *Structure* pose a problem for the rationality of science insofar as there is no neutral way to adjudicate which paradigm is better. The second challenge is the pessimistic induction (PI), whereby the repeated failure throughout the history of science to propose true theories about unobservable entities undermines the rational confidence in the claims of our current scientific theories. Roush rejects both challenges to the rationality of science and argues that we need to pay more detailed attention to the history of science.In particular, she argues that there has been a novelty in the method of science that has not been adequately appreciated, hence the induction of PI is not a good one. From this, she concludes that we should not be persuaded by the pessimistic induction.

In the next chapter, Cyrus M. Mody considers a less-often-discussed, but still integral, aspect of *Structure*: normal science. Mody canvasses a selection of traditions in science studies to argue that Kuhn makes the important discovery that normal science proceeds successfully precisely because it is not primarily concerned with achieving truth or knowledge about nature. Rather, normal science is focused on making science workable. In other words, scientists and engineers are able to achieve the continuation and success of their given paradigm because they are not concerned with specific objections to the paradigm raised in the form of anomalies.Instead, they are driven by a set of ever-shifting aims, which are determined, in part, by their professional communities and societies. Kuhn's analysis of normal science leads Mody to the question, "What do scientists (and engineers) do all day?" Mody examines various sub-categories of normal science, such as "sub-normal science" and "abnormal science," which remain important aspects of the regular work of scientists. Mody concludes that a Kuhnian task that remains today is to ask what scientists and engineers get out of such activities in normal science, as this will help us to better understand the more-often-discussed topics of scientific revolutions and scientific change promulgated in *Structure*.

In Chapter Eight, Rogier De Langhe and Peter Rubbens revisit Kuhn's idea of the "essential tension" between tradition and innovation in science. They suggest that the problem of theory choice concerns finding the right balance between the exploitation of existing theories and the exploration and creation of new ones. The fact that a scientist is confronted with the choice between expanding on an existing theory or starting work on a new one, turns the problem of theory choice under risk to a problem of theory choice under uncertainty, because the number of alternatives is potentially infinite. There is no way to specify the criteria for a good theory in advance. Following Kuhn's approach, De Langhe and Rubbens argue that the solution is to base theory choice on heuristics, rather than on an algorithm. They introduce an agent-based model that shows how a decentralized group of scientists, each following a heuristic of balancing exploitation and exploration, can result in a robust Kuhnian pattern normal science punctuated by periods of revolution.

In Chapter Nine, James Marcum examines Kuhn's evolving notion of incommensurability and its influence on Kuhn's notions of reality and truth. Here, Marcum identifies four distinct versions of the incommensurability thesis: (1) the original thesis of incommensurability (IT_O), which was motivated by Kuhn's "hermeneutical turn" in his attempt to understand Aristotle's *Physics* through Newtonian mechanics; (2) the version of the incommensurability thesis found in the *Postscript* (IT_P), which marks Kuhn's "linguistic turn" towards the thesis as he defended it from the charges of irrationalism and relativism; (3) the local incommensurability thesis (IT_L), introduced by Kuhn (1983) as a response to the charge that his notion of incommensurability is implicitly incoherent; and (4) the taxonomic incommensurability thesis (IT_T), which was motivated by Kuhn's "evolutionary turn" concerning scientific development, and so marked a methodological shift from a historical philosophy of science (HPS) to an evolutionary philosophy of science (EPS). Following this reconstruction of Kuhn's incommensurability thesis, Marcum argues that, given Kuhn's later views of EPS and IT_T, Kuhn is best understood as holding the position of adaptive realism along with a pragmatic theory of truth.

In the tenth chapter, Michela Massimi turns to the issue of whether Kuhn's view in *Structure* supports the position of realism or relativism. Massimi presents two readings of Kuhn that interpret him as a mild and sophisticated realist: (1) Paul Hoyningen-Huene's (1993) Kantian interpretation; and (2) Ron Giere's (2013) 'perspectival realist' interpretation. Massimi raises problems for each of these readings. First, following Alexander Bird, she argues that the Kantian reading faces the *challenge of naturalism* and *the challenge of phenomena*. Second, she maintains that the perspectival realist reading faces the problems of *perspectival truth* and *conceptual relativism*. With these challenges in mind, Massimi offers a solution by introducing *naturalized Kantian kinds* (NKKs) and argues that this reading is able to offer an interpretation of Kuhn as a mild realist, while avoiding the central problems raised against the other two realist readings.

In Chapter Eleven, William J. Devlin examines Kuhn's criticism of truth in science. Particularly, Devlin re-constructs Kuhn's rejection of the traditional account of the correspondence theory of truth as an argument that dismisses our access to the truth about the mind-independent world for epistemic reasons. However, he argues that, by removing truth altogether from the project of science, Kuhn faces, what Devlin calls, the *problem of inconsistency*: Kuhn cannot hold, at the same time, the claims that science does not achieve the truth about the world and that it still achieves knowledge about the world. Devlin maintains that, in order to defend Kuhn's philosophical enterprise, an alternative correspondence theory—*the phenomenal-world correspondence theory of truth*—needs to be introduced. Such a theory of truth allows for science to achieve knowledge of a mind-dependent world while, at the same time, remain consistent with Kuhn's rejection of any notion of truth pertaining to the mind-independent world.

In Chapter Twelve, Brad Wray focuses on the complex relationship between Kuhn's social epistemology and the sociology of science, and the implications of this relationship for Kuhn's legacy in the philosophy of science. Wray notes that although the sociology of science had little influence on *Structure*, the *Structure* greatly influenced certain schools in sociology of science (i.e., the Mertonian tradition and the Strong Programme). Furthermore, he contends that Kuhn himself held that the sociology of science was central to his project in *Structure*: a better understanding of the community structure of science is important for Kuhn's epistemology of science (viz., explaining how epistemic considerations enable scientists to resolve disputes without an appeal to truth). Finally, given Kuhn's epistemic work in the philosophy of science, Wray concludes that Kuhn has provided us with the foundations for building a social epistemology of science.

In the final chapter of the volume, Paul Hoyningen-Huene examines the development of Kuhn's ideas before and after *Structure* through his unpublished work. First, Hoyningen-Huene focuses on the penultimate draft of *Structure* (which he calls '*Proto-Structure*'), where he argues that a major break occurred between the two versions on the grounds that two key features are missing from the earlier draft: (1) the influence and discussion of Wittgenstein's theory of concepts and (2) Kuhn's response to the distinction between the context of discovery and the context of justification. As such, Hoyningen-Huene uproots the assumption that these two features were inspiring to the general composition of *Structure*; instead, they are best construed as late additions to the work, given their absence from *Proto-Structure*. Second, Hoyningen-Huene turns to Kuhn's unfinished book manuscript, *The Plurality of Worlds: An Evolutionary Theory of Scientific Development*. There he shows that Kuhn was developing a novel idea concerning a theory of kind terms. This theory is unique insofar as it is naturalistic, multidisciplinary, and draws from developmental psychology. Furthermore, this new theory is integral to Kuhn's work as it lays the foundations for a proper understanding of incommensurability, along with Kuhn's conception of taxonomies, lexicons, revolutionary developments, and reality. Hoyningen-Huene concludes by noting that in light of this unfinished work, the complete development of Kuhn's philosophy, along with the full consequences of his ideas, remains to be seen.

References

Bird, A. 2012. The structure of scientific revolutions and its significance: An essay review of the fiftieth anniversary edition. *British Journal for The Philosophy of Science* 00 (2012): 1–26.
Hanson, N. R. 1958. *Patterns of discovery*. Cambridge: Cambridge University Press.
Kuhn, T. [1962]1996. *The structure of scientific revolutions*. Chicago: University of Chicago Press.
Kuhn, T. 1974. Second thoughts on paradigms. In *The structure of scientific theories*, ed. F. Suppe, 459–482. Urbana: University of Illinois Press.
Kuhn, T. 1987. What are scientific revolutions? In *The probabilistic revolution, volume I: Ideas in history*, ed. Kruger, L. et al. Cambridge: MIT Press. Reprinted in Kuhn (2000).
Kuhn, T. 1990. The road since structure. *PSA* 1990 2. Reprinted in Kuhn (2000), 90–104.
Kuhn, T. 2000. *The road since structure*, ed. J. Conant and J. Haugeland. Chicago: University of Chicago Press.

Lakatos, I., and A. Musgrave, eds. 1970. *Criticism and the growth of knowledge*. Cambridge: Cambridge University Press.
Masterman, Margaret. 1970. *The nature of a paradigm*, in Lakatos & Musgrave (1970).
Scheffler, Israel. 1967. *Science and subjectivity*. Indianapolis: Bobbs-Merrill Company.
Shapere, Dudley. 1964. The structure of scientific revolutions. *Philosophical Review* 73:383–394.
Shapere, D. 1966. *Meaning and scientific change. Mind and cosmos: Essays in contemporary science and philosophy*. Pittsburgh: University of Pittsburgh Series in the Philosophy of Science III (Pittsburgh).

Chapter 2
Kuhn's *Structure*: A Moment in Modern Naturalism

Steven Shapin

The Structure of Scientific Revolutions (henceforth, *Structure*) is history. That's a matter of course; the book offered a theory of historical change in science; it started out by promising a far-reaching change in how we write the history of science; and the cases that made up much of the empirical content of the book were canonical in the academic history of science. *Structure* is, for all that, an odd exercise in the history of science: it's a historically-informed and historically-framed *theory* of science, and, while philosophers routinely produce that sort of thing, historians do so only rarely. The point was made by the Princeton historian of science, Charles Gillispie (1962, p. 1251), reviewing *Structure* for *Science* magazine in 1962: Thomas Kuhn "is not writing history of science proper. His essay is an argument about the nature of science." And this perhaps explains the fact that, when it appeared a half century ago, the historians didn't really know what to make of it, while the philosophers instantly, if perhaps wrongly, thought they knew exactly what kind of thing it was. It was a theory of science which most philosophers attacked whenever they encountered it, and which, if they didn't encounter it, they might conjure up as an ideal-type enemy. *Structure* was a bête-noir of the philosophy of science— it was seen to deny the role, or even the sufficiency in science, of truth, reason, method, reality, and progress. It dismissed method in favor of social consensus or of inarticulable informal criteria; it challenged the notion that science was a peculiarly open-minded practice; it elevated practice over formal theory, the hand over the head and the community over the free and rational individual knower. It commended the philosophical importance of describing science realistically in its making, rather than as its finished products were enshrined in the textbooks.

The philosophical critics were right. Kuhn was a fine rhetorician and he offered his opponents a series of stick-in-the-mind sound-bites, the take-aways, the things you remember about *Structure* when you can remember almost nothing else. On truth:

S. Shapin (✉)
Department of the History of Science, Harvard University,
Cambridge, MA 02138, USA
e-mail: shapin@fas.harvard.edu

© Springer International Publishing Switzerland 2015   11
W. J. Devlin, A. Bokulich (eds.), *Kuhn's Structure of Scientific Revolutions—50 Years On*,
Boston Studies in the Philosophy and History of Science 311,
DOI 10.1007/978-3-319-13383-6_2

"We may... have to relinquish the notion" that scientific change brings scientists "closer and closer to the truth" (Kuhn 1962, p. 169).On scientific education and the mental habits it fosters: "it is a narrow and rigid education, probably more so than any other except perhaps in orthodox theology" (165). On Scientific Method: what Kuhn famously called paradigms "may be prior to, more binding, and more complete than any set of rules for research that could be unequivocally abstracted from them" (46). On the unity of science: science is "a rather ramshackle structure with little coherence among its various parts" (49). On a distinctive scientific rationality: "As in political revolutions, so in paradigm choice—there is no standard higher than the assent of the relevant community" (93). On the insufficiency of logic in science: we must take seriously "the techniques of persuasive argumentation effective within the quite special groups that constitute the community of science" (161). On progress: accepting *Structure*'s picture of science may make "the phrases 'scientific progress' and even 'scientific objectivity'... come to seem in part redundant" (161).

Those sentiments are remarkable, the more so as they were written not, as some critics supposed, by someone meaning to denigrate or attack science, but by someone who, so far as one can tell, thought that, of course, science was a powerful and reliable cultural practice, perhaps the most powerful and reliable way of knowing the world. How is that possible? The answer points to a second sense in which Kuhn's *Structure* is history. It *belongs* to history; it is a historical object, produced in a historically specific set of circumstances. For all that the ideas in *Structure* continue to influence, inform, and, for many, to irritate and enrage, it emerged from a particular historical conjuncture and one way of understanding it is to take a look at some features of that conjuncture—as Kuhn liked to say, *grosso modo*.

The call for understanding *Structure* as coming from, and making sense in, its specific historical circumstances isn't exactly unique. Indeed, during the celebration of fifty years of *Structure*, historicizing the book has probably been the standard gesture in framing commemorative exercises, especially by identifying 'influences' on the type of project represented by *Structure* or on its central ideas—for example, the influence of Conant's pedagogical project on Kuhn's use of case-studies in *Structure*; the influence of what Joel Isaac has recently termed Harvard's "interstitial academy" on Kuhn's interdisciplinarity; the influence of Kuhn's own strikingly loose educational background on what Isaac called his notable "independence of mind" (Isaac 2012, pp. 31–62, 213);[1] the influence of Michael Polanyi on his deployment of the idea of tacit knowledge; the influence of Bruner on his use of Gestalt psychology; of Wittgenstein on rules and rule-following; of Stanley Cavell on all sorts of things, including the awareness of Wittgenstein and of the under-appreciated role of philosophical aesthetics.

---

[1] Robert Merton similarly pointed to Harvard's "microenvironments," allowing Kuhn, or indeed anyone so placed in the institution, serendipitously to stumble on resources and to acquire perspectives which they might not otherwise encounter (Merton 1977, pp. 76–109; Merton and Barber 2004, pp. 263–266).

Still, there's a kind of historical story about *Structure* that isn't so easily folded into notions of 'influence': this is an account of the *conditions of possibility* of some of the basic sentiments in *Structure*, sentiments that mark this book out from almost everything else previously said about the nature of science and its modes of historical change. Those basic sentiments are the ones represented in the sound-bites about truth, reason, method, reality, and progress, and the social virtues of science. They are, so to speak, the water in which the fish of *Structure* move and have their beings, the environment for the rest of *Structure*'s more specific claims, for example, about incommensurability, anomalies, and crisis. When you read *Structure*, it's the nose in front of your face, the things you tend to forget about when your view is set on finer discriminations. It is the historicity of these sentiments that I want to describe, the dispositional framework of *Structure*, not its fine structure, its historical or philosophical scope, or the validity of its propositions about science.

I call these basic sentiments about science *naturalistic*—where naturalism is opposed to normativity, where the naturalist intention is to describe, interpret, and explain and not to justify, celebrate, or, more rarely, to accuse.[2] My historical claim about *Structure* is very simple: its naturalistic sentiments represent some of the things that are intelligibly sayable about science when the normative and celebratory loads of commentary are lightened or removed. It's not hostility to science that makes these sentiments seem like criticism; it's just the absence of celebration. And that's one reason Kuhn was so mystified by scientists who thought that he had described "normal science" as some form of hack-work ideally to be dispensed with, so puzzled by 1960s student radicals who took it as an exposé of scientific authority, and so upset by philosophers like Imre Lakatos who saw a causative link between those "contemporary religious maniacs ('student revolutionaries')"—and what he called Kuhn's view of scientific consensus as "mob psychology" and "mob rule" (Kuhn 1970, p. 259, 2000, p. 308; Marcum 2005, pp. 74–75; Lakatos 1970, pp. 93, 178). Kuhn did not conceive of naturalism *about* science as criticism *of* science; for him, it had no prescriptive or advisory function. There's no sign that in 1962 he saw the avalanche of criticism coming: *Structure* does not have a defensive tone. And Steve Fuller's (2000) dyspeptic assault on Kuhn is surely right on the point that Kuhn intended nothing remotely like criticism of the status quo, though Fuller set aside as insignificant that Kuhn never intended celebration either.

What was it it was about the particular cultural and political environment from which *Structure* emerged that offered the conditions of possibility for its naturalism? Almost needless to say, this environment is not a sufficient condition for sentiments such as Kuhn's—after all, Kuhn's many critics inhabited much the same macroenvironment—but, if they are not sufficient conditions, and if one must also consider smaller-scale environments offered by Kuhn's institutional settings and

---

[2] "Naturalism" in these matters is, of course, a notoriously disputed notion. Here I use it in a deflationary sense routinely deployed by such sociologists of scientific knowledge as Barry Barnes and David Bloor (Barnes et al. 1996, pp. 3, 106, 173, 182, 185, 202, 208; Bloor 1991, pp. 77–81, 84–106, 177–179), where a naturalistic account of science as it actually proceeds is juxtaposed to its celebration, defense, rational reconstruction, or essentialization.

disciplinary identity (or lack of identity), nevertheless I suggest that it was the new cultural and political place of science in the post-War decades that made the naturalism of *Structure* possible.

With the notable exception of Ludwik Fleck's (1935/1979) neglected work— neglected, that is, by practically everyone but Kuhn before the 1960s—there was in academic writing little unambiguously naturalistic sentiment about the nature of science or its modes of change during the first part of the twentieth century. Science was too precious, and especially too fragile, a flower to be dealt with in an ordinary, matter-of-fact sort of way. What it urgently needed was defense, celebration, and justification—demarcation from intellectual pretenders and lesser breeds. Defense and justification were not just ideologically commended; they presented themselves as intellectually compelling. As David Hollinger (1983) and others have shown, Merton's sociological project was crafted partly to display the liberal, critical, and open condition of science as a social institution and so to hold up the scientific community as a virtuous mirror to totalitarian societies thinking they could interfere with its liberal processes and align science with either Fascist or Communist social agendas. Michael Polanyi's anti-rationalist picture of science (1940, 1946, 1958) was an explicit counter to Marxist rationalist projects which reckoned that science could be enrolled in socially valued planned projects in the same way as technology. Polanyi showed that rationalist accounts were contingently, not logically, attached to the defense of science, and it was that defense, the celebration of science as a unique and powerful form of tacit knowledge, that Polanyi had in view. In philosophy, the epistemological project described by Vienna Circle philosophers like Hans Reichen-bach (1938) admitted what was called the "sociological task" of *describing* scientific conduct as it is and as it was, but identified the peculiar epistemological tasks as the normative work of criticism and advising, and, among some members, displaying the Unity of Science that was deemed essential to its cultural authority (Creath 1996; Galison 1998). Karl Popper (1963) took on the urgent job of addressing and identi-fying the methodological distinctions between authentic science and its illegitimate pretenders. In the history of science, George Sarton (1936) famously insisted that science was culturally unique, that the historian of science was not doing anything like the same sort of thing as the historian of religion, war, politics, or art, and that the history of science should show humankind at its most noble and uplifting.[3] Historians of what was once known as an "internalist" disposition took the writings of Marxist historians as denigration and threat, but the Marxists were celebrating science too, though taking a different view on what science was, what its cultural value consisted in, and the conditions of its historical change (Shapin 1992; Kuhn 1968, 1977). For the Marxists, scientific agendas responded to all sorts of economic and social forces, but the location of science between "base" and "super-structure"

---

[3] Alexandre Koyré's work (1939), aimed at displaying the intellectual coherence and intelligibility of past science, drifted into the consciousness of Anglophone historians during and after the War, and Kuhn's excitement at that project is evident in *Structure* and elsewhere. One can see Koyré's historical sensibilities as naturalistic, but he did not offer a *theory* of science and some of his historian-followers would have been appalled at the very idea.

was contested within Marxist thought. Marxism was itself seen as a science, and that tells you much of what you need to know about the extent to which writers like J. D. Bernal thought of science as an ordinary cultural practice.

The conditions of possibility of naturalism about science in the second part of the twentieth century were framed by changes in its political and economic circumstances. Naturalism in the intellectual view of science followed normalization in its institutional environment. The story of the changing place of science in the political economy of post-War America has now been well told by, among others, Daniel Kevles, Paul Forman, Peter Galison, and David Kaiser, and I have nothing here to add to their accounts. State funding for science exploded: in the mid-1960s, it was reckoned that the U.S. government was then spending more on research and development than the entire Federal budget before Pearl Harbor (Price 1962, p. 1099, 1965, p. 3). Physics blazed the trail to Fort Knox but the range of American sciences that benefited from huge increases in Federal financial support was very large. Vannevar Bush's dream in *Science, the Endless Frontier* (1945/1995) was substantially realized in the National Science Foundation, while the National Institutes of Health expanded its already huge existing support of the biomedical sciences. First the GI Bill and then the National Defense Education Act transformed the scale of graduate training in the sciences and, as Kaiser has shown, altered the substance of physics teaching and research (Kaiser 2002, 2004, 2005). A vocabulary was developed to talk about the value of science and it was a vocabulary that testified to the simultaneous normalization of science and to its immense civic worth. The Steelman Report to the President of 1947 referred to scientists as "an indispensable resource" for all sorts of national "progress" (Steelman 1947, Vol. IV, p. 1). With the outbreak of the Korean War, the rhetoric of "resource" was sharpened: scientists now appeared specifically as "tools of war," "a war commodity" and "a major war asset" that could be "stockpiled" just like "any other essential resource" (Smyth 1951). The argument that fundamental research should be valued and supported because of its contribution to civic, commercial, and military goals was institutionalized in American political economy. And, while the material value attributed to scientific research was, and continues to be, subjected to periodic skepticism and even ridicule, it provided a solid and endurable basis for the institutional security of science.

From the point of view of leaders of the scientific community, enough has never been enough, and lamentations over public "ignorance" of science, over rampant pseudoscience and antiscience, and over dangerous declines in funding never ceased (Gordin 2012). Yet, as Daniel Greenberg and others have noted since the early 1960s, these complaints don't very well describe either the continuing largesse of the State or the durable public esteem in which science has been held in this country through the Cold War and beyond (Greenberg 1967/1999, 2001; Shapin 2007). An occasional blip in funding or admiration is no apocalypse and no amount of hand-wringing could persuade disinterested observers that science was not more securely established than it had ever been.

The point here is not whether science has been well, or even very well, treated since the War; it's that it has been increasingly enfolded into normal political, civic, and commercial institutions. Though many people continue intelligibly to talk of relations between "government and science," "the military and science," and "business and science," in fact it has become difficult to understand the nature of government, of war, or of business without understanding the extent to which they all build science into their quotidian conduct. And the talk of science as a separate and distinct institution—as when we routinely refer to the relations between "science" and "society"—increasingly picks out the decreasing quantum of science that is conducted supposedly "for its own sake" and in institutions that Max Weber assumed were uniquely dedicated to the stewardship of such inquiry.

A way into those structures is through three texts produced a year either side of *Structure*. Two appeared in 1961: the first was President Eisenhower's Farewell Address delivered on January 17, 1961 and the second was a paper titled "Impact of Large-Scale Science on the United States," given as a talk in May 1961, and appearing in *Science* several months later, by the Director of the huge Oak Ridge National Laboratory, Alvin Weinberg. Neither of these texts dealt in any substantial way with scientific practice, scientific method, or with cultural change in science—that is, with the central concerns of *Structure*—but each expressed sentiments that relieve science of the cultural armor which historically protected it from the naturalism central to *Structure*.

Two phrases are about all that's commonly remembered from the two 1961 pieces—from the Farewell Address, the coining of the tag "military industrial complex" and, from Weinberg's text, "Big Science," a phrase which was not in fact wholly original. The pieces emerged, with Eisenhower, from the Heart of Political and Military Power and, with Weinberg, from the Heart of Science. And the remarkable thing is that they were *critical* of aspects of science—Big Science, Weinberg suggested, was "ruining science"; scientists were "spending money instead of thought" (Weinberg 1961, pp. 161–162)—and, more to the point, they were *fearful* of it. Science, they said, had grown great, powerful, politically secure, and politically influential. The post-War institutional successes of Big Science had immeasurably enhanced the resources for doing science while they had endangered its integrity and lured science into political arenas in which it historically had no legitimate place. The seventeenth-century Royal Society had committed itself not to "meddle" with "affairs of Church and State," while Eisenhower warned that its current "meddling" threatened the very nature of the democratic order that so recently Merton and others saw as the internal guarantee of its intellectual authenticity and the external guarantee of its institutional existence.[4]

---

[4] Eisenhower noted (1961/1972, p. 207) that the organization of science had experienced a "revolution": the traditional individualistic picture of a "solitary inventor, tinkering in his shop" had quite recently been replaced by "task forces" of scientists, lavishly funded by government contracts and orientated not to the search for truth but to securing even more money to pay for even more expensive equipment. The American scientific community was shocked both at this depiction of their institutional circumstances and at the idea that they should be thought so powerful, and Eisenhower's scientific advisor George Kistiakowsky (1961; see also Price 1965, p. 11) had to reassure them that Eisenhower really meant only to criticize military-orientated research.

The shift from science perceived as delicate to science perceived—at least by some influential commentators—as powerful, even too powerful, was rapid (Agar 2008). In the same year that *Structure* was published, the political scientist Don Price at Harvard wrote (1962, p. 1099) of "the plain fact ... that science has become the major Establishment in the American political system," and a survey of scientists' involvement in nuclear weapons policy by the Princeton political economist Robert Gilpin noted that "The American scientist has become a man of power to perhaps a greater degree than scientists themselves appreciate." In no other nation, "nor in any other historical period, have scientists had an influence in political life comparable to that exercised by American scientists," (Gilpin 1962, p. 299). The reviewer of Gilpin's book in *Science* magazine agreed that, until Hiroshima, "nobody would have dreamed of writing a book on [scientists'] political influence," for they had none (Rabinowitch 1962, p. 974). The points at issue here are not whether these perceptions were either accurate or novel. Criticisms of scientific expertise were *not* unprecedented or global; Eisenhower had quite specifically in mind the activities of such scientist-politicians as Edward Teller and Wernher von Braun (York 1995, p. 147); and what Weinberg meant by Big Science did not describe the institutional environment in which all, or even most, American scientists did their work. Yet these criticisms were targeted at the commanding heights—the most visible sectors— of post-War science and they were articulated from within the corridors of power. Indeed, the most pertinent thing about these views is that they were credible, that they were sayable at all.

The third text, appearing the year after *Structure*, is the now neglected *Little Science, Big Science* by the sociologist of science Derek de Solla Price (1963/1986). Price, like Kuhn, offered not just a theory of science but a wide-focus view of its mode of historical change. As in *Structure*, this was a theory of science wholly disengaged from celebration or justification. Differences between Price's and Kuhn's enterprise are obvious: science for Price was a unity, while for Kuhn it was an unruly collection of practices each regulated by its own paradigm; Price treated science as a black-box, sucking in quantifiable inputs (scientific practitioners, financial resources, instruments) and generating quantifiable outputs (publications, discoveries, more scientists), while for Kuhn science was, again, an assemblage of conceptual and instrumental projects. Science for Price was no special thing, standing outside of history: Price aimed at, and thought he had achieved, a science of science, establishing that scientific growth could be understood as a natural phenomenon, displaying a "common natural law of growth." All elements of science grew exponentially, but there were others things in society that grew in similar ways. If the doubling period for scientific outputs was fifteen or twenty years, about the same period obtained for such non-scientific things as the Gross National Product and the increase in college entrants per thousand of population. In that sense, science *was* progressive but not uniquely so. Even the sense of remarkable acceleration in scientific growth since the War was normalized in Price's account: in fact, science had *always* grown at the rate seen in the past generation; it was *always* modern, *always* seeming to stand outside of history. The only thing that one might identify as historically novel about present circumstances was that this long-standing rate of growth was about to reach "saturation": you could not have more scientists than there were people, more funds for science than the GNP, and that inflection point in the logistic curve was now visible just over the horizon. Yet, in this academic idiom so different from Kuhn's, Price's enterprise also naturalized and normalized science, and in that respect it was also a sign of its times.

The institutional, economic, and political circumstances of Big Science in the Cold War decades formed the conditions of possibility for *Structure*'s naturalism, but this is not the same thing as saying that naturalism about science was normal in that setting or that justificatory and celebratory sensibilities did not continue to flourish. Academic disciplines do respond to their contexts, but they usually do so in mediated ways, shaped by long-established evaluative traditions, and maybe Kuhn reflected Cold War conditions of complacency about science so well just because he was, in the best sense of the word, a great amateur, not formally trained in, and not securely belonging to, *any* of the academic disciplines concerned with talking about the nature of science.

Structure's naturalism, in the event, was precarious and unstable, and one mark of that precariousness appeared in subsequent work by Kuhn himself. After *Structure*, and especially after the hostile 1965 London conference whose proceedings were published as *Criticism and the Growth of Knowledge*, Kuhn (1970) was cautious about repeating the naturalistic sentiments quoted at the beginning of this piece. He defended *Structure*, of course, but he devoted much energy to specifying just how those naturalistic sentiments should *not* be understood. "'But I didn't say that! But I didn't say that! But I didn't say that!'" Kuhn found himself repeatedly insisting, especially in response to irritating misreadings by student radicals who saw the paradigm concept as evidence of "oppression," but more subtly with respect to academics made anxious by the naturalistic sentiments of *Structure* (Kuhn 2000, p. 308). The last chapter of *Structure*, the 1969 "Postscript" to the second edition, and subsequent essays, all testify to Kuhn's anxieties. There must, he thought, be ways of talking legitimately about scientific progress, about scientific truth, about the moral and procedural specialness of scientific communities, and, of course, there must be a way to produce a historically robust theory of science while avoiding odious relativism. He knew that *Structure* had exploded the usual supports for ideas of scientific progress, rationality, and realism, so new ones should be found.

Late in his life, Kuhn observed that "I haven't produced any children." He greatly admired his students John Heilbron and Paul Forman, but said that both had "turned entirely away from" the sort of history of science that he did, and that only Jed Buchwald, an under-graduate, not a graduate, student of Kuhn, did the close analysis of scientific ideas with which Kuhn identified his own historical work (Kuhn 2000, pp. 304, 319). But Kuhn *did* have intellectual offspring, and his reaction to those children is further evidence of his reflective ambivalence towards the naturalism of *Structure*.

The scholars who not only found Kuhn's naturalism congenial but who enthusiastically incorporated aspects of it into substantive sociological and historical work were, of course, my former colleagues at the Edinburgh Science Studies Unit— Barry Barnes and David Bloor— and associated sociologists in England, including Michael Mulkay, Harry Collins, and Trevor Pinch. Bloor (1976/1991) understood the "Strong Programme" in the sociology of knowledge as a form of Kuhnian naturalism and Barnes's book *T. S. Kuhn and Social Science* applauded *Structure* as "one of the few fundamental contributions to the sociology of knowledge" (Barnes 1982, p. x). To my knowledge, Kuhn never commented on the substance of any of this work, but his overall assessment is well known: addressing Harvard's Department of the History of Science in 1991, he announced that all of it was "deconstruction gone mad," a judgment which soon went viral among the anti-relativist warriors in the science wars of the 1990s (Kuhn 2000, p. 110). The point is not whether Kuhn disowned his intellectual progeny for good reasons—in my view, his account of this work was unfortunately quite wrong—rather, it's one index among many of how fragile naturalism about science was and continues to be.

That's because the institutional and cultural normalization of science that was the condition of possibility for *Structure*'s naturalism was never complete, not in the culture as a whole and only partially in the academic disciplines concerned with the nature of science and its history. The science wars were one sign of this patchy normalization; the fetishization of Scientific Method in the contemporary human sciences is another. Here again, the history of science is much more than a topic of inquiry for the academic discipline of the same name. For instance, the scientific naturalists of the Victorian era thought that the march of progress would inevitably deliver a secularized culture, science triumphant over religion. They were wrong about the religion bit, but they could not visualize the institutional and civic security of science a hundred years on.

What about the stories historians of science tell themselves about their own field? In recent times, we have become very good at debunking teleologically progressivist narratives about science, and, in that debunking, Kuhn has been a hero. (After all, that's how *Structure* begins, with a promise to deliver history from the myth-tellers.) But historians have not been keen to see themselves and their work as historical objects. Rejecting simple-minded stories about *scientific* progress, we tend to take for granted that the *historical* stories we now tell about science are so obviously better than they used to be, and we lack curiosity about the circumstances that have made those stories possible. Kuhn's *Structure* was a moment in modern naturalism, not a rung on the ladder of inevitable historical progress. Its conditions of possibility include the institutional state of science in the post-War decades; its conditions of fragility include the only partly normalized institutional and cultural state of science today.

References

Agar, J. 2008. What happened in the sixties? *The British Journal for the History of Science* 41:567–600.
Barnes, B. 1982. *T. S. Kuhn and social science*. London: Macmillan.
Barnes, B., D. Bloor, and J. Henry. 1996. *Scientific knowledge: A sociological analysis*. Chicago: University of Chicago Press.
Bloor, D. 1976/1991. *Knowledge and social imagery*. 2nd ed. Chicago: University of Chicago Press.

Bush, V. 1945/1995. Science—the endless frontier: A report to the President on a program for postwar scientific research. National Science Foundation 40th anniversary edition. Washington, DC: National Science Foundation.

Creath, R. 1996. The unity of science: Carnap, Neurath, and beyond. In *The disunity of science: Boundaries, contexts, and power,* ed. P. Galison and D. J. Stump, 158–169. Stanford: Stanford University Press.

Eisenhower, D. D. 1961/1972. Farewell address. In *The military-industrial complex,* ed. C. W. Pursell Jr., 204–208. New York: Harper and Row.

Fleck, L. 1935/1979. *Genesis and development of a scientific fact.* Chicago: University of Chicago Press.

Fuller, S. 2000. *Thomas Kuhn: A philosophical history for our times.* Chicago: University of Chicago Press.

Galison, P. 1998. The Americanization of unity. *Daedalus* 1998 (Winter): 45–72.

Gillispie, C. C. 1962. The nature of science. *Science* 13:1251–1253.

Gilpin, R. 1962. *American scientists and nuclear weapons policy.* Princeton: Princeton University Press.

Gordin, M. 2012. *The pseudoscience wars: Immanuel Velikovsky and the birth of the modern fringe.* Chicago: University of Chicago Press.

Greenberg, D. S. 1967/1999. *The politics of pure science.* 2nd ed. Chicago: University of Chicago Press.

Greenberg, D. S. 2001. *Science, money, and politics: Political triumph and ethical erosion.* Chicago: University of Chicago Press.

Hollinger, D. A. 1983. The defense of democracy and Robert K. Merton's formulation of the scientific ethos. In *Knowledge and society,* ed. R. A. Jones and H. Kuklick, Vol. 4, 1–15. Greenwich: JAI Press.

Isaac, J. 2012. *Working knowledge: Making the human sciences from Parsons to Kuhn.* Cambridge: Harvard University Press.

Kaiser, D. 2002. Scientific manpower, Cold War requisitions, and the production of American physicists after World War II. *Historical Studies in the Physical and Biological Sciences* 30:131–159.

Kaiser, D. 2004. The postwar suburbanization of American Physics. *American Quarterly* 56:851–888.

Kaiser, D. 2005. *Drawing theories apart: The dispersion of Feynman diagrams in postwar physics.* Chicago: University of Chicago Press.

Kistiakowsky, G. 1961. Quoted in G. DuShane, Footnote to history. *Science* 133:355.

Koyré, A. 1939. *Etudes galiléennes.* 3 Vols. Paris: Hermann.

Kuhn, T. S. 1962. *The structure of scientific revolutions.* Chicago: University of Chicago Press.

Kuhn, T. S. 1968/1977. The history of science. In *The essential tension: Selected studies of scientific tradition and change,* ed. T. S. Kuhn, 105–126. Chicago: University of Chicago Press.

Kuhn, T. S. 1970. Reflection on my critics. In *Criticism and the growth of knowledge,* ed. I. Lakatos and A. Musgrave, 231–278. Cambridge: Cambridge University Press.

Kuhn, T. S. 2000. *The road since structure: Philosophical essays, 1970–1993, with an autobiographical interview,* ed. J. Conant and J. Haugeland. Chicago: University of Chicago Press.

Lakatos, I. 1970. Falsification and the methodology of research programmes. In *Criticism and the growth of knowledge,* ed. I. Lakatos and A. Musgrave, 91–196. Cambridge: Cambridge University Press.

Marcum, J. A. 2005. *Thomas Kuhn's revolution: An historical philosophy of science.* London: Continuum.

Merton, R. K. 1977. The sociology of science: An episodic memoir. In *The sociology of science in Europe,* ed. R. K. Merton and J. Gaston, 3–141. Carbondale: Southern Illinois University Press.

Merton, R. K., and E. Barber. 2004. *The travels and adventures of serendipity.* Princeton: Princeton University Press.

Polanyi, M. 1940. The rights and duties of science. In *The contempt of freedom: The Russian experiment and after,* ed. M. Polanyi, 1–26. London: Watts & Co.

Polanyi, M. 1946. The foundations of freedom in science. *Bull Atomic Scientists* 2 (11–12): 6–7.

Polanyi, M. 1958. *Personal knowledge: Towards a post-critical philosophy*. Chicago: University of Chicago Press.

Popper, K. R. 1963. *Conjectures and refutations: The growth of scientific knowledge*. London: Routledge and Kegan Paul.

Price, D. de S. 1963/1986. *Little science, big science... and beyond*. New York: Columbia University Press.

Price, D. K. 1962. The scientific establishment. *Science* 136:1099–1106.

Price, D. K. 1965. *The scientific estate*. Cambridge: Harvard University Press.

Rabinowitch, E. 1962. Scientists and politics [review of Gilpin 1962]. *Science* 136:974–977.

Reichenbach, H. 1938. The three tasks of epistemology. In *Experience and prediction: An analysis of the foundations and structure of knowledge*, ed. H. Reichenbach, 3–15. Chicago: University of Chicago Press.

Sarton, G. 1936. *The study of the history of science*. Cambridge: Harvard University Press.

Shapin, S. 1992. Discipline and bounding: The history and sociology of science as seen through the externalism-internalism debate. *History of Science* 30:333–369.

Shapin, S. 2007. Science and the modern world. In *The handbook of science and technology studies,* ed. E. Hackett, O. Amsterdamska, M. Lynch, and J. Wajcman, 3rd ed. 433–448. Cambridge: MIT Press.

Smyth, H. D. 1951. The stockpiling and rationing of scientific manpower. *Bull Atomic Scientists* 7 (2): 38–42, 64.

Steelman, J. R. 1947. Science and public policy: A report to the President by John R. Steelman, chairman, The President's Scientific Research Board, 5 Vols. Washington, DC: Government Printing Office.

Weinberg, A. M. 1961. Impact of large-scale science on the United States. *Science* 134:161–164.

York, H. F. 1995. *Arms and the physicist*. Woodbury: American Institute of Physics.

Chapter 3
Kuhn and the Historiography of Science

Alexander Bird

3.1 Introduction

A useful way to understand Thomas Kuhn's thinking about the nature of the history of science is to see it as embodying two principal strands of historicism, both of which we can find in G.W.F. Hegel. While Kuhn certainly did not acquire his views about the history of science from thinking about Hegel on history, brief reflection on Hegel's historicism will help illuminate analogous elements in Kuhn's thought. Doing so will bring us to a better understanding of the relationship between Kuhn's conception of the history of science and that promoted by many more recent students of science studies, in particular exponents of the Sociology of Scientific Knowledge (SSK). In the light of his historicism, we can see why Kuhn took a predominantly internalist approach to the explanation of scientific change whereas SSK adopted externalism.[1]

3.2 Hegelian Strands in Historicist Historiography

Hegel's view of the relationship between philosophy and its history, as articulated in his *Lectures on the History of Philosophy* (Hegel 1825), is often summarised as claiming that philosophy is the history of philosophy. Hegel contrasts his approach to philosophy with the ahistorical approach typical of the enlightenment. Descartes, for example, sought a method for philosophy and for the foundations of science

---

[1] This chapter expands on ideas first presented in Bird (2012b) and discussed at a meeting of the Institute of Historical Research, to whom I am grateful for helpful comments.

A. Bird (✉)
Department of Philosophy, University of Bristol, Cotham House,
Cotham Hill, BS6 6JL Bristol, UK
e-mail: alexander.bird@bristol.ac.uk

© Springer International Publishing Switzerland 2015
W. J. Devlin, A. Bokulich (eds.), *Kuhn's Structure of Scientific Revolutions—50 Years On*,
Boston Studies in the Philosophy and History of Science 311,
DOI 10.1007/978-3-319-13383-6_3

that would be a valid method in any context of enquiry and would yield results of permanent value. Similarly, Kant sought to develop an ethics from principles of pure reason; again, both the validity of the method and the truth of its deliverances were intended to be sempiternal. For both Descartes and Kant, and the majority of the philosophers of the enlightenment (and indeed before and since), if a philosophy is to be satisfactory, both its methods and its results should stand for all time, independently of the particular historical conditions in which they were produced. According to Hegel, a philosophical idea or argument can be neither understood nor (therefore) evaluated independently of the historical context in which it is produced. To engage with philosophy is necessarily therefore to engage with its history. (This historicist approach to philosophy and to ideas in general is also to be found in the work of Giambattista Vico and of Hegel's immediate predecessor, Johann Gottfried von Herder.)

As I have just articulated it, Hegel's historicism may seem to imply both relativism and contingentism, that there are no absolute truths since all truth is relative to an historical context and that there are no general explanations in history, even in history of philosophy, since ideas and actions are the product of local rather than general factors. What makes Hegel's historicism interesting is that it denied both of these. On the contrary, according to Hegel there are important absolute truths and values, and furthermore these play a role in explaining the historical development of ideas. For Hegel, the Absolute Idea, or World Spirit, determines the evolution of history: "History is the Idea clothing itself with the form of events" (Hegel 1821, § 346). That is, there is an underlying 'logic' to history, from which a pattern emerges in historical evolution.

Thus, there may appear to be a tension within Hegel's historicism, as both implying and rejecting relativism—perhaps encapsulated by the tag 'objective idealism' used to describe Hegel's philosophy. But on closer inspection this tension disappears. First, the relativism of the first kind really concerns the rationality of agents and societies. It is in the interpretation and evaluation of what people say and do that we need to refer to their historical context. *Methodologically*, it might also make sense to treat truth as relative: in assessing the genius of Ptolemy's *Almagest* it is not appropriate to criticise him on the grounds of being factually wrong in his geocentrism. This relativism is entirely consistent with the claim that underlying the evolution of human history is a process that has a certain direction and structure or mechanism. There is an analogy here with Kant's distinction between the phenomena (which are relative to the subject's forms of intuition and categories) and the noumena (which are absolute).[2] Indeed, it is the very evolution of history that gives rise to

---

[2] There are of course important differences. Hegel's Absolute is part of the world of ideas, whereas Kantian noumena are not. Kant denies that the noumena are knowable, whereas Hegel does claim to know something about the Absolute. The latter leads to a familiar point on which there is indeed a tension: how can the relativist rationally make any claims about general and absolute truths? Hegel is alive to this point, even if he does not resolve it entirely. On the one hand, the philosophical historian seeks an insight into the abstract reason that lies behind historical processes. On the other hand, no historian can avoid some element of subjectivity, for that is essential to historical interpretation.

the different eras of thought in relation to which particular ideas must be evaluated. So, in fact, the two elements of Hegel's historicism are connected. It is *because* thought or consciousness evolves that we need to consider ideas in their historical context.[3] Frederick Beiser distinguishes a horizontal level in Hegel's philosophy of history from a vertical level (Beiser 1993, pp. 279–280). The former concerns a society's or nation's particular circumstances (e.g. geographic or demographic), and the uniqueness of these means that there is no comparison of societies against an absolute standard; we can assess them only relative to those peculiar circumstances. But there is also a vertical level, that of the development of world history, and nations can be judged according to their contribution to progress towards the end of history, the self-consciousness of freedom.

The first dimension of Hegel's historicism, that which is concerned with tradition and understanding ideas from their historical context, I call the *conservative* strand of historicism, In so doing, I employ Karl Mannheim's contrast between 'conservative' thinkers who emphasise the importance of tradition and history and those who endorse an Enlightenment 'natural law ideology' (Mannheim 1953).[4] The second dimension, which identifies laws or patterns in the development of history, I call the *determinist* strand. This strand is exemplified also by Karl Marx and by Auguste Comte, and is the dimension of historicism attacked by Karl Popper (Popper 1957). According to determinist historicism, the historian is not limited to describing and explaining particular events but may hope also to see in the many particular events an underlying pattern. In this respect, history has one affinity with the sciences. The historian of genius, like a great scientist, will not only identify such patterns but may also seek to explain those patterns by reference to an underlying mechanism.

3.3 Historicism in the Work of Kuhn

3.3.1 Conservative Historicism in Kuhn

Kuhn's approach to the history of science exemplifies both strands of historicism identified in the work of Hegel.[5] From the perspective of the philosopher of science,

---

A reflective, philosophical historian is aware of this and so may avoid merely imposing their preconceptions on the historical data, which is the danger facing an historian seeking general laws but who is unaware of the historical conditioned nature of their own thought. By being aware both of the contextual nature of thought and of the existence, albeit obscured, of an underlying reason, the philosophical historian can partially transcend his or her own era. But only partially, which is why Hegel (1821) tells us "The owl of Minerva spreads its wings only with the falling of the dusk", intimating that it is only with the end of history that we can be in a position to understand fully how the Absolute shaped the unfolding of history.

[3] To give an anachronistic analogy: it is because life-forms have evolved that, to understand the nature of a species, we need to understand the ecological environment in which that species originated. (This analogy stands despite the important difference between biological evolution and Hegel's historical evolution in that the latter has a teleological aspect that the former lacks.)

[4] See Bloor (1997) for more on understanding Kuhn in the light of Mannheim.

[5] Reynolds (1999) also mentions Kuhn in connection with different species of historicism.

the conservative strand is the most striking. For the logical empiricist and positivist philosophy of science that preceded Kuhn held that the evaluation of a scientific theory is *sub specie aeternitatis*; theory assessment is a matter of applying the laws of theory confirmation to the total available evidence.[6] Those laws should tell one how well the theory is supported by that evidence. And those laws, like the laws of deductive logic, hold for all people at all times. Kuhn's radical departure from this application of 'natural law ideology' to science was to suggest that the evaluation of a theory is relative to a specific tradition of puzzle-solving.[7] Kuhn regarded the term 'paradigm' as having two senses (Kuhn 1970, pp. 174–175). The broader sense encompasses the shared commitments of a scientific community, for which he also used the term 'disciplinary matrix' (Kuhn 1970, p. 182). The second sense is more narrow, and refers to the most central of the community's commitments, its *exemplars* (Kuhn 1970, p. 187). Exemplars are the community's exemplary solutions to its scientific puzzles. These set the standards for subsequent science in that domain. A piece of good science, a satisfactory proposed solution to a puzzle, will resemble an exemplar puzzle-solution. Hence theory evaluation is not context-independent, but is relative to a puzzle-solving tradition. Furthermore, the puzzles themselves emerge from the puzzle-solving tradition. Worthwhile puzzles are ones that resemble existing puzzles or emerge from gaps in the existing tradition. So the Newtonian tradition sets the puzzles of reconciling the observations of planets, satellites, and comets to Newton's laws and of measuring the value of the gravitational constant G (among many others). The importance of tradition shows itself also in the phenomenon of incommensurability.[8] Kuhn claims that a scientific theory may not even make sense to someone working outside the tradition from which it originates (Kuhn 1970, pp. 149–150). There may be incomplete understanding because not only is evaluation anchored in the exemplars of the tradition, but so also are important elements of the meanings of scientific terms.

3.3.2   Determinist Historicism in Kuhn

From the logical empiricist perspective, the history of science should show no interesting patterns. The evolution of modern science is the story of discoveries building

---

[6] While there was disagreement about what those laws were, or whether they should in fact be laws of falsification, logical empiricists agreed that such laws would be perfectly general and context-independent.

[7] The idea of tradition is central to Kuhn's description of normal science and the function of paradigms (see for example Kuhn 1959, p. 227; Kuhn 1962, p. 10). Hegel also refers to the importance of tradition in science, "likewise, in science, and specially in Philosophy, do we owe what we are to the tradition which, as Herder has put it, like a holy chain, runs through all that was transient, and has therefore passed away" (Hegel 1825, pp. 2–3).

[8] Beiser says that, as portrayed by Hegel, the values of each nation and the manners in which they achieve the self-awareness of freedom are incommensurable between nations (Beiser 1993, pp. 279–280).

on and adding to the stock of pre-existing knowledge. Armed with a general scientific method and logic of confirmation, science will inevitably accumulate discovered truths. The process may be accelerated thanks to scientists of genius in the right place at the right time, it being acknowledged that the process of discovery (as opposed to justification) cannot be entirely methodical. But deviation from the accumulation of knowledge is rare, being mostly due to (often externally driven) deviations from proper methods (e.g. Lysenkoism); in some cases, a false theory might look initially attractive (e.g. the caloric theory of heat), but the growing weight of evidence would eventually point in the right direction.

Against this expectation of accumulation, Kuhn's claim that there is a much more interesting, fundamentally cyclical pattern with its alternating phases of normal and extraordinary (revolutionary), science represents a significant challenge. In comparison to logical empiricism, it underplays progress. If true, it would suggest that there are *systematic* deviations from the method and logic of science. The transition from normal to revolutionary science, says Kuhn, proceeds via a crisis. Revolutions are messy affairs in which the change of paradigm is contested. The existence of regular, seemingly inevitable crises and revolutions indicates that scientific progress cannot in fact be the accumulation of truths. The fractious nature of revolutions suggests that scientists cannot be applying general rules of confirmation. Thus, the idea that there is a pattern to scientific change, a law of scientific development is an important component of Kuhn's thought and represents a radical break with the preceding orthodoxy. In passing we may note the parallels between Kuhn's account of change in scientific ideas and Hegel's account of the transformation of the Absolute Spirit (Bird 2000, pp. 129–130). In the latter, a thesis gives rise to a second idea, the antithesis, in conflict with the first, the conflict being resolved in the synthesis; likewise, in the former, research within a paradigm generates an anomaly, leading to resolution through revolutionary change, where the new paradigm retains elements of its predecessor while creatively accommodating the anomaly that caused the revolution.

3.3.3   The Two Strands of Historicism United in Kuhn

I mentioned that the two strands of historicism are linked in Hegel's work. The same is true for Kuhn, though in an importantly different way. In Hegel's case, it is because of the historical development of the Absolute that ideas are historically conditioned. So, determinist historicism implies conservative historicism. And to an extent we may say the same about Kuhn, for if there are radical, incommensurable breaks in scientific thought, then the assessment of a scientific idea will require placing it in its correct scientific era.

There is however another, deeper link between the conservative and determinist strands in Kuhn's historicist historiography of science, operating in the other direction, from the conservative to the determinist. As mentioned, scientists may identify patterns in the phenomena; they may also wish to explain those patterns by reference to underlying mechanisms or more general laws. Kepler identified the elliptical nature of the orbits of the planets, and other patterns besides; Newton explained these with his theory of gravitation. Mendeleev discovered the periodic pattern among the elements; this was explained by the atomic theory developed by Rutherford, Bohr, and Chadwick. To see a pattern in the history of science is one thing, to explain it is another, although in reality these two processes are not so easily separated. The determinist strand in Kuhn's thinking gives us his belief that there is a cyclical pattern in the history of science. The conservative strand, the fact that science evolves by exploiting a paradigm-based tradition of puzzle-solving, explains the pattern.

Normal science exists because a scientific field is dominated by a set of exemplars. As mentioned above, these exemplars set the agenda for future research, such as showing how all objects in the solar system conform to Newton's laws (in the Newtonian paradigm). Not only were these problems made relevant by Newton's *Principia*, but the latter also furnished the means of solving those problems, primarily though examples of using the theory to solve problems of this sort. This explains the existence of normal science. Not all normal science puzzle solving is straightforward. For example, mathematical astronomers in the eighteenth century found it difficult to reconcile the observed orbit of the moon with Newton's theory. Alexis Clairaut and Jean d'Alembert calculated the value for the period of revolution of the Moon's perigee, which is the point on the Moon's orbit that is nearest the Earth. This they found to be eighteen years, though it was known from observation to be half that. Such apparent conflicts between the observed phenomena and theory, along with other cases where scientists fail to solve puzzles, are *anomalies*. Kuhn explains that anomalies are not regarded as counter-evidence against the theory at the centre of the paradigm (Kuhn 1970, p. 80). During normal science, the failure to solve a puzzle is attributed to the scientist or the scientific community. But if anomalies accumulate that are particularly significant and recalcitrant, then the blame for the anomalies may begin to shift away from the scientists and towards the paradigm. This is what occurs during the periods of crisis. The anomalous motion of the moon was sufficiently serious for Leonhard Euler to suggest that Newton's law of gravitation might need adjusting, until Clairaut showed that the anomaly was due largely to the inaccurate approximations being used. This might be thought of as a mini-crisis that was successfully resolved within the paradigm. More serious was the crisis that arose in the late nineteenth century stemming from the anomalous precession of the perihelion of Mercury, reported by Urbain Le Verrier in 1859, and arguably the null outcome of the Michelson–Morley experiment. Since, according to Kuhn's conservatism, normal science requires an established tradition with a credible paradigm, crises must be resolved. If they are not resolved within the existing paradigm, then that paradigm must be replaced. In particular, it must be replaced by a paradigm that can support a puzzle-solving tradition. Thus we have scientific revolutions. In this way, the conservative strand in Kuhn's historicism (the emphasis on a tradition of puzzle-solving) explains the determinist strand (the law-like cyclical pattern of scientific change).

3.4   Kuhn's Internalist Historiography of Science

Kuhn's work was a stimulus to a wide range of science studies from history of science through to sociology of science, and many of those working in these fields see themselves as, in a loose way, heirs to a Kuhnian legacy. Barry Barnes's book *T. S. Kuhn and Social Science* is just one prominent example of this. Kuhn (1992) himself nonetheless repudiated in the strongest terms the most important (and philosophically most sophisticated) school within science studies, the Strong Programme in the Sociology of Scientific Knowledge (SSK), of which Barnes was a leading light. Furthermore, the scope of Kuhn's criticisms embraced implicitly a broader spectrum of the science studies movement than Barnes's Edinburgh School alone.

In this section, I wish to explain in what way Kuhn rejected the social constructivism found in much SSK, and why he was right to do so in the light of his commitment to historicism. The social constructivism that Kuhn found antithetical to his own ideas holds that the principal factors in determining the outcome of a scientific episode, such as a crisis, are social and political factors originating outside science—what came to be known as *externalist* history and sociology of science.[9] According to this approach, the triumph of Pasteur's theories rejecting spontaneous generation is not the result of the probative power of his experiments with swan-necked flasks. Rather, that success may be attributed to the fact that his ideas were better attuned to the views of the conservative, Catholic hierarchy that ruled in the France of Louis Napoleon (Farley and Geison 1974; Farley 1978). The success of Darwinism is not held to be a consequence of the arguments presented in the *Origin of Species*, but is instead to be explained by the natural sympathy of free-market Britain to the idea that improvement is the outcome of unfettered competition (Young 1969). These are examples of external explanations of scientific change, contrasting with internalist explanations that refer only to aims, values, practices, and beliefs originating within science.

Kuhn's own account of science leaves little room for such external influences, certainly not enough for them to be the principal determinants of the outcomes of scientific debates. Let us first consider normal science. As explained, progress during normal science is driven by the paradigm, the set of exemplary puzzle-solutions that define a puzzle-solving tradition. These set the agenda—they define what kinds of puzzles are worth pursuing and they set the standards by which proposed solutions to those puzzles are assessed. Kuhn's account leaves no space for external influences in either regard. Since, as Kuhn emphasizes, the bulk of scientific activity is normal science, it follows that at least most scientific change is governed by factors internal to science.

Kuhn explicitly endorses a predominantly (but not exclusively) internalist approach. He tells us that "... the ambient intellectual milieu reacts on the theoretical

---

[9] The internalism versus externalism debate is perhaps somewhat outdated now. Yet it was very much alive in Kuhn's lifetime and his work gave an impetus to it—often in a manner of which he disapproved.

structure of a science only to the extent that it can be made relevant to the concrete technical problems with which the practitioners of the field engage" and goes on to criticize historians (from outside history of science) who ignore this fact (Kuhn 1971, pp. 137–138). Kuhn acknowledges that (typically older) history of science could be limited by exclusive internalism, but "that limitation," he says, "need not always have been a defect, for the mature sciences are regularly more insulated from the external climate, at least of ideas, than are other creative fields" (Kuhn 1971, p. 148–149). To the general insulation of scientific ideas from external influences, Kuhn makes two exceptions. While the development of a tradition is internally driven, the *origins* of that tradition need not be: "Early in the development of a new field... social needs and values are a major determinant of the problems on which its practitioners concentrate" (Kuhn 1968, p. 118). Kuhn contrasts this with the later evolution of a scientific specialty, "The problems on which such specialists work are no longer presented by external society but by an internal challenge to increase the scope and precision of the fit between existing theory and nature" (Kuhn 1968, p. 119).[10]

Kuhn's second exception concerns the *rate* at which science develops. Kuhn tells us that the timing of a scientific advance can be conditioned by external factors (Kuhn 1968, p. 119). That must be correct, if only because prevailing economic conditions will determine the quantity of resources put into research. Kuhn also suggests that because the various scientific disciplines interact, there may be a cumulative effect from external factors on the evolution of science. Advances in technology clearly make a difference to the ability of science to progress.

It is important to note that neither of Kuhn's two exceptions suggest that external influences regularly influence the *outcome* of a scientific investigation or debate. In SSK, one can distinguish a *weak* programme, which looks at the broad social and political environment and its effect on, for example, the existence of scientific institutions, as exemplified by Merton's "Science, technology and society in seventeenth century England" (Merton 1938), and a *strong* programme, as exemplified by Shapin and Schaffer's *Leviathan and the Air-Pump* (Shapin and Schaffer 1985), according to which the content of the accepted results of science and the very terms of scientific discourse are influenced by social and political factors. At most Kuhn's work gives partial support to the weak programme. For example, concerning the crisis in Ptolemaic astronomy preceding the Copernican revolution, Kuhn tells us that one ingredient is "the social pressure for calendar reform, a pressure that made the puzzle of precession particularly urgent." He goes on to tell us, concerning a scientific crisis, that "In a mature science—and astronomy had become that in antiquity—external factors like those cited above are principally significant in determining the timing of

---

[10] I note that Kuhn's internalism does mark an element of difference from Hegel's conservative historicism. For the latter is justified in part by a holism about thought. According to Hegel, the various components of a society, from its politics and religion to its culture and philosophy form an inseparable whole. And so when one element changes so do all including its philosophy (Hegel 1861; c.f. Beiser 1993, p. 274). One would naturally take Hegel to include science in this. Kuhn, however, claims that a modern science is largely insulated from external changes, its origins and pace of progress excepted.

breakdown, the ease with which it can be recognized, and the area in which, because it is given particular attention, the breakdown first occurs." While acknowledging that such factors can be important, Kuhn emphasises that "technical breakdown would still remain the core of the crisis" (Kuhn 1970, p. 69). So while external factors may influence the manner in which this episode occurs, it remains the case that internal factors explain why it could occur at all.[11]

Even if normal science and crisis can be explained on purely internal grounds, perhaps we might expect externalism to be more likely to be true of Kuhn's account of revolutionary science? Kuhn himself writes:

> Individual scientists embrace a new paradigm for all sorts of reasons and usually for several at once. Some of these reasons—for example, the sun worship that helped make Kepler a Copernican—lie outside the apparent sphere of science altogether. Others may depend upon idiosyncrasies of autobiography and personality. Even the nationality or the prior reputation of the innovator and his teachers can sometimes play a significant role. (Kuhn 1962, pp. 152–153)

One should not over-emphasize the externalism of even this passage. As Kuhn says, *some* of the reasons an individual has for adopting a paradigm may lie outside of science, and he gives only one example, Kepler's sun-worship. By implication, he thinks that the 'others' he mentions do not lie outside of science. Clearly reputation is internal to science. It is true that differences in personality might make scientists differ in the degree to which they are disposed to adopt radical ideas. Or a scientist's openness to an idea might be influenced by the fact that she was trained in a laboratory where such ideas were developed, or because working on that theory offers better job prospects. But again it is not clear that these are considerations external to the practice of science, at least not in a way that threatens the central internalist claim that is important to Kuhn—that it is the requirement of puzzle-solving that overwhelmingly determines which ideas are developed and adopted. As Kuhn says immediately after the quoted passage, "Probably the single most prevalent claim advanced by the proponents of a new paradigm is that they can solve the problems that have led the old one to a crisis. When it can legitimately be made, this claim is often the most effective one possible" (Kuhn 1962, p. 153). The most effective way of advancing a new paradigm is to show that it solves the problems that led the old one into crisis. Kuhn then goes on to point out that this may not always be possible; indeed, the new candidate paradigm may not help at all with the crisis-evoking problems. In that case, novel predictions, predictions of phenomena that would be entirely unsuspected under the old paradigm, can be persuasive (such as the prediction of the phases of Venus by Copernicus's theory). Kuhn then mentions the role of aesthetic considerations. He also discusses at length the characteristic of revolutions we have subsequently called 'Kuhn-loss' and the importance of a new paradigm being a fruitful basis for new problem-solving research. In assessing whether Kuhn gave direct encouragement to externalist study of science, we must set the short

---

[11] For more detail on Kuhn's internalism in relation to SSK, see Bird (2012a).

quoted passage against the six pages that follow, in which he emphasizes in detail the importance of the puzzle-solving tradition in determining its own development.

During revolutionary science, however, it might be thought that there is no puzzle-solving tradition to provide this determining role. And this would suggest that extra-scientific forces may be able to fill the gap and swing the outcome, as Barnes (1981; 1990) argues. This, I believe, is a misinterpretation of Kuhn that is based on the idea that revolutions are all-encompassing and radical breaks with the past. While it is true that Kuhn may have overstated the difference between normal and revolutionary science, it is also true that Kuhn lays great emphasis on progress and continuity through revolutions. The final chapter of *The Structure of Scientific Revolutions* is entitled "Progress through Revolutions". The penultimate chapter, "The Resolution of Revolutions", describes the constraints imposed on the new paradigm by the long-standing success of its predecessor in puzzle-solving. Such constraints mean that there is significant continuity in revolutionary science.[12] From the perspective of the internalism/externalism debate, there is more in common between normal and revolutionary science than there are differences. In both normal and revolutionary science, the principal driver of progress is scientific problem solving. During normal science, the need to solve problems is satisfied by the paradigm. During extraordinary science, the need remains, but now must be satisfied by finding a replacement paradigm. What determines the outcome will still be, above all, the power of a proposed paradigm to solve puzzles. That may not determine the outcome uniquely and unambiguously—Kuhn stresses that there is room for rational disagreement about the relative problem-solving power of a proposed new paradigm compared to the old one or a competitor. Nonetheless, the fact that the dispute is about scientific puzzle-solving power restricts the choices available. The puzzle-solving tradition thereby still exerts its force during revolutionary science, though not in as straightforward or as decisive a way as during normal science. The participants in the debate must be able rationally to believe that their favoured solution will deliver more and better puzzle solutions than its competitors. In particular, supporters of a new paradigm must, in most cases, be able to show that it solves a sizeable proportion of the most significant anomalies that beset the old paradigm, while also preserving the bulk of the puzzle-solving power of its predecessor. Since finding an innovative solution that achieves this is not easy, most episodes in revolutionary science will provide very few choices. Typically, there will be only a single revolutionary proposal to challenge the old paradigm. Given the infinite range of beliefs a scientist could have about a given subject matter, all but a small handful are straightforwardly excluded by factors internal to science, even during extraordinary science.

---

[12] Kuhn returned to this theme in several of his later writings, for example in his essay "Objectivity, Value Judgment, and Theory Choice" (Kuhn 1977b), in which he articulates the five scientific values (accuracy, consistency, breadth of scope, simplicity, fruitfulness). His concern throughout is to reject accusations of subjectivity in theory preference, while allowing space for reasonable disagreement across paradigms.

Of course, this does seem still to leave some room for external factors to influence the outcome of a scientific revolution. Nonetheless, I do not think that it was Kuhn's view that such factors play a determining role. The fact that there is room for rational disagreement does not mean that the view of any individual scientist, let alone the views of the community as a whole, must be swayed by something else. What it does mean is that the resolution of a revolution will be a much more protracted affair. Within normal science there may be disputes, but typically these can be resolved using the resources of the paradigm. The causes of AIDS were initially disputed, but standard techniques identified a particular virus as the cause in a way that is beyond rational dispute. In such a case, there is no Kuhn-loss—no pre-existing beliefs and commitments need to be given up; the success of the viral explanation is clear by established standards; and the research opportunities (and so scientific benefits) afforded by the new discovery are transparent.

On the other hand, in revolutionary science there is Kuhn-loss to be weighed against the claims of new puzzle-solving power; there are at least some conflicts with existing standards or beliefs; and because of this, the potential for the alleged new discovery to support a fruitful programme of research (future puzzle-solving) is unclear, especially when we have to give up an existing tradition. When Barry Marshall and Robin Warren proposed that the principal cause of gastric ulcers is a bacterium rather than, as had been believed, excess acid brought about by factors such as stress, a whole sub-field of research (as well as treatment) was under threat; consequently it was unclear at the time whether, in puzzle-solving terms, the new proposal would be more productive than the established view. We are not comparing like with like in such a case, because we are comparing an existing track-record with future promise. So there is plenty of room for difference of opinion as to whether the new view should be adopted or not. Biographical factors, as Kuhn says, may play a part in determining how individual researchers respond. Older scientists will have kudos, expertise, research programmes and laboratories invested in the established approach whereas younger scientists will see the newer view as offering them opportunities for speedier advancement than they might otherwise have had. But such room for differences of opinion and influence by professional considerations cannot persist for ever. As time goes by, the puzzle-solving power of the new view will turn from potentiality to actuality and a more direct comparison of old and new will be possible. In the case of Marshall and Warren, after initial resistance, community opinion did concur reasonably speedily. While there may be no definitive point at which it becomes irrational to stick with one or other view, that does not mean that it is reasonable to hold on to either view indefinitely. Although one can find scientists who continued to believe in the luminiferous aether in the 1920s, most theoretical physicists had accepted Einstein's (special) theory of relativity, originating in 1905, well before the outbreak of World War I. At the same time, the advantages that may attract an ambitious young scientist to a new field will soon tarnish if it fails to live up to its promise as a vehicle for a productive puzzle-solving tradition; cold-fusion is a case in point. So the difference between normal and extraordinary science is not one between phases when internal or external factors are decisive. It is a difference in the speed and manner in which internal factors, unaided, reach their conclusion.

A different reason for thinking that external factors must be significant is the thought that the questions upon which scientists work are very frequently determined by the material needs of broader society. Bacon's vision in the *Novum Organum* is for a scientific enterprise that leads to economic prosperity and so one would expect such science to be concerned with questions directly connected with problems arising in the social and economic sphere. One might think of the effort made by astronomers to solve the problem of longitude in this vein. Given Kuhn's insistence on the insulation of mature science from external sources of problems, it is not surprising that Kuhn makes an explicit differentiation between science and technology. "As a first approximation," he says, "the historian of socioeconomic development would do well to treat science and technology as radically distinct enterprises, not unlike the sciences and the arts" (Kuhn 1971, p. 143). Technology does respond to external demands, but science does not. Of course, that may be a mere definitional distinction, but Kuhn gives us reason to think that it is not (Kuhn 1971, p. 142, 147). For, he argues, science and technology have historically been distinct spheres of activity. It was only in the middle third of the nineteenth century that Bacon's vision began to be achieved, and knowledge generated by scientists began to make a technological difference to society, first through dyestuffs and then later through electrical devices and techniques such as the pasteurization of beer, wine, and milk. Of course, the insulation of science from technology is not guaranteed, and one might wonder whether modern science is in a different position. And certainly the requirements of governments for research to respond to externally generated challenges may blur the distinction between science and technology to the point of eradication. Whatever the truth may be about the practice of science today, Kuhn's view of the distinction between science and technology aligns with the difference of the source of the problems—internal and external, respectively—that motivate the intellectual activity in each.

3.5   Historicism and Internalism

Above I noted a *prima facie* tension between the relativism implied by Hegel's conservative historicism and the objectivism implied by his determinism. Likewise it would appear that Kuhn's emphasis on paradigms (disciplinary matrices, tradition) implies relativism whereas the rejection of externalism, implied by his determinism, is associated with objectivism about scientific knowledge. This tension is only apparent in both cases. I briefly discussed Hegel's attempted resolution above: it is the deterministic evolution of the Spirit that generates the different stages of development in which ideas are anchored; the philosophical historian can transcend this to some extent, if not completely, to see the working of reason in the Absolute. Because Kuhn's direction of explanation is the other way around, from the existence of traditions (conservatism) to the pattern of normal science and revolution (determinism), Kuhn's emphasis on relativism is somewhat stronger than Hegel's.

In brief, the answer in Kuhn's case is this: while externalism leads to relativism (or scepticism), the reverse is not the case—relativism does not necessarily lead to externalism. It is true that objectivists, those who believe that science has reasonable success in uncovering facts about an independent world, will be internalists. But it does not at all follow from this that all internalists must be objectivists. Internalism makes room for both objectivists and relativists who believe that the determinants of scientific change are encapsulated within science itself. And that is the kind of internalist I take Kuhn to be.

Indeed, Kuhn has to be this kind of internalist if he is to be true to the determinist strand of his historicism. If externalism were true, so that factors originating outside science were the main drivers of scientific change, then there would be no reason to suppose that there would be any patterns in the history of that change. Instead, one would expect the history of science to demonstrate the same chaos and contingency that we find elsewhere in human affairs. Take the two examples from nineteenth century biology I mentioned earlier—Pasteur's rejection of spontaneous generation and Darwin's account of evolution through natural selection. If the externalists are right, then the inception and success of these two theories are the results of socio-political forces that happen have opposing natures, occurring at the same time in different countries: clerical conservatism in France, economic liberalism in Britain (note that one of the alleged political advantages of Pasteur's results is that they challenged atheistical Darwinism, which many held to require some form of spontaneous generation). Since these different social forces are the products of different sequences of historical events in the two countries, it is difficult to see how the totality of forces such as these could conspire to produce the orderly cycle that Kuhn sees in the history of science. To use a mechanical analogy, Kepler was able to discern the elliptical orderliness of the solar system because the solar system is a simple and isolated system. If it were frequently perturbed by large inter-stellar objects passing through or nearby, then there would have been no elliptical orbits for Kepler to discover. Likewise, a necessary condition of the truth of Kuhn's theory is that the drivers of (the content of) scientific development are local to science, which is largely isolated from the influence of other developments in history.

Externalism and determinist historicism in science could be reconciled if the patterns in the history of science reflected larger patterns in history that pervade the social and political as well as the scientific. The laws of scientific development would be manifestations of a broader historicist truth. Yet this seems implausible for two reasons. First, such global historical determinism has little credit. The great schemes of Hegel and Comte are believed by few, if any, today, and not even all Marxists accept a strict determinism along the lines, for example, of Lenin's version of dialectical materialism. Second, such external historical determinism must *explain* the Kuhnian cyclical pattern. No attempt has been made to show how such an explanation would proceed. Indeed, it seems implausible that there could be any such explanation. For historical determinists tend to see history as exhibiting large-scale stages (Comte, Marx), but those could not explain Kuhn's cyclical pattern (especially as there is not one such pattern, but many, as the pattern for one field of science need not coincide with the pattern for another field). Furthermore, historical determinists tend to believe that history has a direction, even a goal (Hegel). Not only does Kuhn deny that the history of science has a direction, but to accept that it does would be to permit a kind of whig history of science that SSK rejects (Kuhn 1962, p. 172).[13] So, even if (implausibly) an external determinism could account for Kuhnian patterns, this would be antithetical to the kinds of externalism promoted by many practitioners of science studies.

3.6   History of Science and Philosophy of Science

"History, if viewed as a repository for more than anecdote or chronology, could produce a decisive transformation in the image of science by which we are now possessed" (Kuhn 1962, p. 1). The image Kuhn has in mind is partly historical but is primarily philosophical. Kuhn's historicism, I argue, makes an important contribution to meeting his philosophical aims. Kuhn's philosophical target was logical empiricism. The logical empiricists, construed broadly enough to include Popper, were concerned to give normative accounts of theory change. History of science can be used to test these normative accounts—on the assumption that scientists normally reason as they ought to reason. This last assumption is important for without it the normative theory could be a theory about how scientists ought to change their reasoning habits in order to improve them. One can see Bacon's philosophy of science in that light. As it was, the logical empiricists did believe that scientists reason correctly, by and large; their theories therefore aimed to articulate how scientists do in fact reason. Popper not only thought that scientists should reject theories that are falsified, he also held that they do in fact reject such theories. So, Popper's view would itself face falsification if the history of science could show that scientists regularly do hold onto theories in the face of apparently contradictory evidence. This is indeed what Kuhn aims to show with the conservative component of his historicism, according to which normal science is governed by a puzzle-solving tradition. As we have discussed, in Kuhn's view, scientists do not abandon a tradition in the face of an anomaly. Rather, an anomaly will often be just another puzzle to solve. If a scientist tackles such a puzzle but fails to solve it, that failure is attributed to the limitations of the scientist, not of the tradition. So the very existence of normal science is a major challenge to Popper's falsificationism.

    Things are somewhat different with respect to the more central inductivist strand of logical empiricism. Here Kuhn's target is the conception of science as an accumulation of true beliefs acquired by the repeated application of the scientific method (e.g. some form of inductive logic). Such a view is consistent with the existence of normal science. It is revolutionary science that creates the problem for logical empiricism, for these are episodes when well-established beliefs are rejected. However, since such episodes are, in Kuhn's terminology, 'extraordinary', there is room for debate as regards their evidential value against the logical empiricist picture. For their

---

[13] Note that whiggism is a feature of Marxist historians.

relative rarity will allow the logical empiricist to regard them as occasional exceptions, in some cases pathological episodes (or corrections to pathological science) or features of immature science, and so forth. This is where the determinist strand of Kuhn's historicism becomes relevant. Since Kuhn can show that scientific change has structure, the cyclical structure of normal science–crisis–revolution–normal science, then non-cumulative episodes, revolutions, cannot be dismissed in this way; they are inevitable parts of the scientific process.

Kuhn himself aims for a major revolution in the philosophy of science. He rejects the common assumptions of the logical empiricists that the aim of science is truth and that scientific rationality consists in applying some kind of logic to the relationship between a theory and straightforward assertions concerning the scientist's experience. Kuhn's replacement paradigm is intended to be one in which the aim of science is puzzle-solving and scientific rationality consists in matching proposed puzzle-solutions to exemplary puzzle-solutions.[14] Kuhn's view need not seem quite so utterly radical when we consider that much human cognition takes place via pattern recognition (think of face recognition) (*cf.* Margolis 1987). However, in his own historical context, the proposal was radical and was perceived as more extreme than it ought to have been. For it was taken as a form of irrationalism about science. Once perceived in that light, it is no surprise that Kuhn's detractors and supporters alike took Kuhn to be articulating a vision of science in which scientists and their ideas, unconstrained by rationality, are subject to social forces.

I have argued that Kuhn's view of science was precisely *not* this. Kuhn wishes to argue for his reconceived understanding of scientific rationality precisely by pointing to the pattern he perceives in the history of science, for the latter is explained by that reconception of rationality better than by the old conception. Thus it is the conservative strand of his historicism that supports that reconception of rationality. At the same time, the rationality of science according to either conception requires determinism. A significantly externalist component in science would undermine the deterministic strand of Kuhn's historicism, and so is antithetical to Kuhn's philosophical aims for the history of science. Thus, I hope that thinking of Kuhn as a historicist regarding the history of science will allow us to rethink his understanding of that subject and, in turn, will allow us to get a better perspective on his philosophical aims concerning the nature and rationality of science.

References

Barnes, B. 1981. On the 'hows' and 'whys' of cultural change (response to Woolgar). *Social Studies of Science* 11:481–498.
Barnes, B. 1982. *T. S. Kuhn and social science*. London: Macmillan.

---

[14] See Bird (2005) for details. This view is contentious in that Kuhn did not proclaim himself as seeking to revise our notion of rationality. That is because the very notion of rationality is close to the idea of following rules of reason. Kuhn showed that science dispensed with rules but instead employed reasoning by analogy with exemplars. The latter, he was at pains to emphasize, is not in any way irrational.

Barnes, B. 1990. Sociological theories of scientific knowledge. In *Companion to the history of modern science,* ed. R. C. Olby, G. N. Cantor, J. R. R. Christie, and M. J. S. Hodge, 60–73. London: Routledge.

Beiser, F. C. 1993. Hegel's historicism. In *The Cambridge companion to Hegel,* ed. F. C. Beiser, 270–300. Cambridge: Cambridge University Press.

Bird, A. 2000. *Thomas Kuhn.* Chesham: Acumen.

Bird, A. 2005. Naturalizing Kuhn. *Proceedings of the Aristotelian Society* 105:109–127.

Bird, A. 2012a. Kuhn, naturalism, and the social study of science. In *Kuhn's The structure of scientific revolutions revisited,* ed. V. Kindi and T. Arabatzis, 205–230. New York: Routledge.

Bird, A. 2012b. La filosofía de la historia de la ciencia de Thomas Kuhn. *Discusiones Filosóficas* 13:167–185.

Bloor, D. 1997. The conservative constructivist. *History of the Human Sciences* 10:123–125.

Farley, J. 1978. The social, political, and religious background to the work of Louis Pasteur. *Annual Reviews in Microbiology* 32:143–154.

Farley, J., and G. Geison 1974. Science, politics and spontaneous generation in nineteenth-century France: The Pasteur–Pouchet debate. *Bulletin of the History of Medicine* 48:161–198.

Hegel, G. W. F. 1821 *(German edition: Grunilinien der Philosophie des Rechts). The philosophy of right,* English edition, T. M. Knox, ed., trans. (1967). London: Oxford University Press.

Hegel, G. W. F. 1825 *(German edition: Vorlesungen über die Geschichte der Philosophie). Lectures on the history of philosophy,* English edition, E. S. Haldane and F. H. Simson, eds., trans., (1955). London: Routledge and Kegan Paul.

Hegel, G. W. F. 1837 *(German edition: Vorlesungen über die Philosophie der Weltgeschichte). Lectures on the philosophy of history,* English edition, H. B. Nisbet, ed., trans., (1975). Cambridge: Cambridge University Press.

Kuhn, T. S. 1959. The essential tension: tradition and innovation in scientific research. In *The third (1959) university of Utah research conference on the identification of scientific talent,* ed. C. Taylor, 162–174. Salt Lake: University of Utah Press. (Page references to the reprint in T. S. Kuhn (1977a)).

Kuhn, T. S. 1962. *The structure of scientific revolutions.* Chicago: University of Chicago Press.

Kuhn, T. S. 1968. The history of science. In *International encyclopedia of the social sciences,* vol. 14, ed. D. L. Sills, 74–83. New York: Crowell Collier and Macmillan. (Page references to the reprint in Kuhn (1977)).

Kuhn, T. S. 1970. *The structure of scientific revolutions.* 2nd ed. Chicago: University of Chicago Press.

Kuhn, T. S. 1971. The relations between history and the history of science. *Daedalus* 100:271–304. (Page references to the reprint in T. S. Kuhn (1977a)).

Kuhn, T. S. 1977a. *The essential tension.* Chicago: University of Chicago Press.

Kuhn, T. S. 1977b. Objectivity, value judgment, and theory choice. In *The essential tension,* ed. T. S. Kuhn, 320–339 Chicago: University of Chicago Press.

Kuhn, T. S. 1992. The trouble with the historical philosophy of science. *Robert and maurine rothschild distinguished lecture 19 November 1991. An occasional publication of the department of the history of science.* Cambridge: Harvard University Press.

Mannheim, K. 1953. *Essays on sociology and social psychology.* London: Routledge and Kegan Paul.

Margolis, H. 1987. *Patterns, thinking, and cognition. A theory of judgment.* Chicago: University of Chicago Press.

Merton, R. K. 1938. Science, technology and society in seventeenth century England. *Osiris* 4:360–632.

Popper, K. 1957. *The poverty of historicism.* London: Routledge and Kegan Paul.

Reynolds, A. 1999. What is historicism? *International Studies in the Philosophy of Science* 13:275–287.

Shapin, S., and S. Schaffer. 1985. *Leviathan and the Air-Pump: Hobbes, Boyle, and the experimental life.* Princeton: Princeton University Press.

Young, R. 1969. Malthus and the evolutionists: the common context of biological and social theory. *Past and Present* 43:109–145.

Chapter 4
From Troubled Marriage to Uneasy Colocation: Thomas Kuhn, Epistemological Revolutions, Romantic Narratives, and History and Philosophy of Science

Alan Richardson

> *There have been philosophers of science, usually those of a vaguely neo-Kantian cast, from whom historians can still learn a great deal.*
> —(Thomas Kuhn 1977b, p. 11)

4.1 A Beginning

If it was still possible to think a new thought about Thomas Kuhn's work before the Jubilee Year 2012, is it still possible today? That is a difficult question to answer, and so, in proper scholarly fashion, I will set it aside. I make no claims about the newness of any of the considerations I engage in below. I do think they are worth thinking again even if we have indeed thought them before.

4.2 A Second Beginning

This essay began life as a short set of remarks for the joint plenary session on Kuhn for the History of Science Society and Philosophy of Science Association meetings in San Diego in November 2012. As I spent that fall trying to figure out what to say, I came increasingly clearly to the conclusion that I should begin with advice to

---

This essay's first incarnation was commissioned by the chairs of the PSA and HSS 2012 program committees, Andrea Woody, Janet Browne, David Kaiser, for the plenary session on Kuhn at the collocated meetings that fall. I wish to thank them and also Alisa Bokulich and William J. Devlin for attempting to get me to think new thoughts about Kuhn.

A. Richardson (✉)
Department of Philosophy, The University of British Columbia,
Vancouver, BC, Canada
e-mail: alan.richardson@ubc.ca

© Springer International Publishing Switzerland 2015
W. J. Devlin, A. Bokulich (eds.), *Kuhn's Structure of Scientific Revolutions—50 Years On*,
Boston Studies in the Philosophy and History of Science 311,
DOI 10.1007/978-3-319-13383-6_4

younger scholars. Here it is: if someone offers you the opportunity to speak briefly about the one figure about whom everyone in your proposed audience has strong and settled opinions, think at least twice before you say yes. But that was the task set for me: to speak about Thomas Kuhn. I was happy to express my deep admiration for his work, but I lacked the fruits of significant, new scholarly labour and, thus, unable to tell my audience much of anything about his work they did not already know.

Faced with this situation, it seemed to me important to shift the focus from what I was doing in speaking at the session, to the larger question of what we all were doing there at that session. Why do these societies meet in the same place and the same time every even-numbered year and why had the organizers gone to the extra-ordinary lengths it must have taken (I have been PSA program chair, I know these things) to set up a joint plenary on Kuhn? I was able, therefore, to fulfill a dream I have had for at least a quarter of a century and ask a large group of assembled historians and philosophers of science "what are we doing here?" This question could be further specified in the case of Kuhn: what goal might an almost unprecedented plenary session of our two societies have, when that session is organized around the work of someone who must be acknowledged as one of the great figures in both history and philosophy of science, but whose legacy is not universally viewed as *grosso modo* positive, certainly not among philosophers of science?

The question has some poignancy. Every second year, the colocated meetings of HSS and PSA serve, as much as anything, as a sort of epitome or emblem of the under-performance of that late 1950s and 1960s intellectual formation titled "history and philosophy of science." There is surprisingly limited traffic between the two societies. The programs are generally put together with very little interchange between the two societies—often you have to choose between important events (should I go to the PSA Presidential Reception or the HSS distinguished lecture?); presumably, this is because neither society expects the choices raised by these clashes to be hard for the average member of their society to make. The feel of these colocated meetings is, frankly, not one of an exciting joint project historians and philosophers of science or even of an intriguing difference of opinion leading to productive engagement. Instead, it feels like if the philosophers can offload much of the organizational work to historians, we can get better appetizers at receptions: shrimp is for PSA; stale pretzels are for APA.

Back in the 1970s a number of philosophers of science—Ron Giere, Ernan Mc-Mullin, Dick Burian, and others—wondered out loud about "the marriage between history and philosophy of science" and whether it was a good or, indeed, a really possible thing (Giere 1973; McMullin 1976; Burian 1977; compare Giere's recent retrospective, Giere 2011). It is hard to imagine an odder metaphor to work within than this one, whether elaborated in a religious or secular way, when considering either discipline formation or interdisciplinary research. With very few exceptions—such as Mary Domski's and Michael Dickson's likening of Michael Friedman's work to a marital aid in the title of their Festschrift for him (Domski and Dickson 2010)—this talk has gone away. In an essay almost baroque in its allusions to romantic literature, Lorraine Daston (Daston 2009) recently has suggested that the dalliance with science and technology studies among historians of science might well be over and that maybe they could find their old flames in philosophy of science again on Facebook. Following Daston's lead, but with a continued lowered tone, we might, using the parlance of the twenty-first century, say that historians and philosophers of science are colleagues with potential benefits, benefits in which, however, we generally lack interest and of which we thus do not avail ourselves.

Kuhn's relations to the project of history and philosophy of science at the time of its institutionalization are indeed interesting. One could well argue that *Structure* made history and philosophy of science in America—the burgeoning project of history and philosophy of science acquired its blockbuster, a book that everyone read and that made history and philosophy of science a project nearly everyone in the academy knew at least a little bit about and respected at least a little bit. Of course, as Kuhn himself later reflected (Kuhn 2000, p. 308), the student radicals of the late 1960s misread the book as an exposé of the authoritarian workings of scientific power. It was, instead, a description of authority and dogma within science: the motor of scientific progress was puzzle solving, and authority and dogma were among the faces by which puzzle solving traditions took hold, were elaborated and advanced, and ultimately broke down. However much Kuhn made history and philosophy of science, it could be argued that history and philosophy of science made Kuhn: after all, the—to use McMullin's phrase—*annus mirabilis* of history and philosophy of science in America was 1960 (the year of the founding of the departments at Pitt and Indiana and of the Center for Philosophy and History of Science at Boston University).[1] By the time Kuhn's book came out, there was an institutionalized project in at least several prominent places that was ready and able to take it seriously and was ripe to discuss the relations between history and philosophy of science as elaborated in Kuhn's book.

So, if we wished to speak the language of a certain branch of science and technology studies, we could talk of the co-construction of Kuhn's intellectual prominence and the project of American history and philosophy of science. But, however that may be, Kuhn himself was famously and pointedly dubious about certain models of history and philosophy of science. His 1968 lecture on the topic given at Michigan State (then revised and published in *The Essential Tension*; Kuhn 1977b) is a long meditation, written, interestingly, mainly from the point of view of graduate training and reflecting his professorial experience in the history and philosophy of science classroom. In it, he posited a difference in interests, goals, reading habits, and so on between the historically-minded and the philosophically-minded among his students. In Kuhn's view, the budding philosophers in his classes were interested in the universal and the normative—they were interested in speaking the truth about what is always and everywhere the same in science and in evaluating specific scientific interventions (say Cartesian physics or Darwinian biology) in light of those universals. The budding historians were interested in the time-bound and local aspects of the texts in front of them. This all sounds very familiar to those of us in the field.

---

[1] This phrase was in his talk at the fiftieth anniversary session of the Boston Center for Philosophy and History of Science in October 2010.

Kuhn ends this essay with a resounding rejection of the marriage plot and a more puzzling (given what he had said throughout the essay) invocation of active discourse (Kuhn 1977b, p. 20): "I urge that history and philosophy of science continue as separate disciplines. What is needed is less likely to be produced by marriage than by active discourse."

Kuhn's sensibilities about a difference of goals and interests between history of science and philosophy of science did not in fact arise from the hard experience trying to give courses to both history and philosophy students at Princeton—however much we can empathize with Kuhn's plight there. They were well-forged already in 1962 and are illustrated in the differences between Mary Hesse's review of Kuhn's *Structure* (Kuhn 1962/2012) and Kuhn's own review of Hesse's 1961 book, *Forces and Fields* (Hesse 1961). Hesse's book was concerned with the development of the field concept against the background of historical arguments about action at a distance. Hesse was one of the first—in her *Isis* review of *Structure* in 1963—to use Kuhnian terminology to frame the significance of Kuhn's own book. The essay concludes:

> It cannot be disputed that this is the first attempt for a long time to bring historical insights to bear on the philosophers' account of science, and whatever the puzzles are that remain to be solved, Kuhn has at least outlined a new epistemological paradigm which promises to resolve some of the crises currently troubling empiricist philosophy of science. Its consequences will be far-reaching. (Hesse 1963, p. 287)

Here, we see the familiar—and surely correct—view that *Structure*, by disrupting "the image of science by which we are now possessed" (Kuhn 1962/2012, p 1), disrupts received philosophy of science which, after all, is the place, according to *Structure,* that that vision is most diligently articulated.

Interestingly, Kuhn's own *American Scientist* review of Hesse's book (Kuhn 1962) takes a very different approach. Kuhn praises the book for both its historical and philosophical concerns—and, in so doing, takes for granted that we, as readers of Hesse's book, know how to sort her concerns into the historical and the philosophical. After noting that Hesse explicitly frames her history of the conceptual development of the field as a criticism of logical empiricist doctrine, Kuhn writes the following:

> The explicitly philosophical dimension of Dr. Hesse's book is for me at least as rewarding as the explicitly historical, but the attempt to combine the two is a source of the book's main weakness. History and philosophy of science can do much to illuminate and stimulate one another. Dr. Hesse's book is a noteworthy attempt to prove the point. Yet the writing of history differs from that of philosophy in its main concerns, values, and principles of organization, and Dr. Hesse has not always succeeded in reconciling the two. (Kuhn 1962, p. 442A)

So the very Kuhn who sought to alter the ground of philosophy of science by offering "a role for history" of science that would disrupt and replace the "image of science by which we are now possessed" himself was able in his reading practices to keep the concerns, values, and principles of organization of history and philosophy of science distinct. Moreover, and importantly, Kuhn specifies that it is in the *writing* of history and of philosophy where these differences are exhibited.

I do not mean to present that as a cosmic mystery or a paradox or an objection to Kuhn. I think one can begin to make sense of what Kuhn is on about here. What is interesting is that, as we attempt to interpret Kuhn on this matter, it will turn out that the "role for history" in *Structure* is more subtle than what most philosophers think the "historical turn in philosophy of science" was suggesting. Here's a conjecture regarding Kuhn's reading of Hesse, Hesse's reading of Kuhn, and then one about Kuhn's understanding of Kuhn. (These are conjectures—I make no strong historical claims regarding their accuracy, but I am interested in their consequences regardless of their truth. I have, by Kuhn's lights, philosophical concerns here.)

Here's how I think Kuhn is reading Hesse's book. Hesse begins (Chap. 1: "The Logical Status of Theories") with a summary of then current logical empiricist accounts of scientific theory and of bridge principles as the keys to the application of theory to the world of experience or ordinary objects. She notes several in principle objections to such views (based largely on model theoretic considerations); she then introduces her own favoured machinery for theorizing the application of theory to observable objects—machinery involving metaphor, analogy, and various types of models. She then goes to the history of the field concept to both develop and illustrate the machinery. I think she reads Kuhn's book the same way: the machinery of logical empiricism is shown to be insufficient to account for scientific development and so we need the machinery of paradigms, normal science, and so on.

But Kuhn does not read his own book that way. He does not develop in principle objections to logical empiricism, introduce new machinery, and then illustrate both the problems for logical empiricism and the solutions to those problems offered by his own machinery through historical examples—at least that is not the primary use of history in the book. Rather, he begins with historiographic questions: what sorts of questions does the working historian need to answer to provide a coherent and explanatory narrative of some episode in the history of science. Then he notices that those questions are not the questions that logical empiricist philosophy of science and, say, the discovery/justification distinction embedded within it would have suggested needed to be answered. The primary lesson is not that logical empiricism is wrong about science but that it gives the wrong advice on what questions to answer to the historian of science. Organized along logical empiricist principles, a history of science is a baffling set of questions without answers, or with answers that do not explain what happened and why in the development of science.

The set of historical facts of the development of science is not, on this view, primarily a source of evidence for or against a philosophy of science, or for illustration of the workings of old or new philosophical machinery. Rather, history of science as a practice engaged in by historians demands the formation of coherent and explanatory historical narratives and the practices involved in the creation of those narratives themselves demand answers to different sorts of questions than the default philosophical machinery would lead you to ask in the first place. In somewhat different words: For Kuhn, the practice of history—the development of historical understanding itself—stands in complicated but ultimately incompatible relations to the sorts of concerns and the machinery for understanding science posited in the logical empiricist philosophy of science (as he understood it) of the 1950s. Or in a(n anti-)Lakatosian slogan, one might say: For Kuhn, history of science as an academic discipline constructs rational reconstructions of the development of science in the sense that it must yield a coherent and explanatory account of the facts, but this practice operates on different principles from those found in the philosophy of science of his day.

I think that this conjecture is interesting because, if it is right (or in the vicinity of right), then it disrupts some widely-circulating stories about the history and philosophy of science. For example, one such story about what I called above the under-performance of history and philosophy of science is that, actually, it turned out to be pretty easy and not terribly interesting for philosophers to do what the "historical turn" asked them to. Philosophers of science, with either a few apologies for, or explanations of, why they had been talking about towers and shadows or black ravens and white shoes and not about the development of scientific knowledge, could acknowledge that of course philosophy of science needed to account for the facts of that development. Those facts are evidence or subject matter or evaluative matter for philosophy of science. In the main, the task of philosophy of science proceeds as before, with a minor clarification of its relations to history. On the view scouted above that is not a sufficient response to Kuhn's position, which was about an incompatibility of the processes of historical and philosophical understanding of science.[2]

Notwithstanding my waving off of the question of truth, I do not think I am wholly off-base in this type of conjecture about Kuhn's self-understanding. To illustrate this, allow me to return to the Kuhn essay with which I started (Kuhn 1977b) and quote from the same final paragraph of it I quoted before, a passage that might seem to fit oddly with both philosophers' understandings of themselves and lessons historians have taken from Kuhn (anti-rational reconstructionist stories):

> About these . . . processes [by which theories develop and are evaluated] we know very little, and we shall not learn more until we learn properly to reconstruct selected theories of the past. As of today, the people taught to do that job are historians, and not philosophers. (Kuhn 1977b, p. 30)

Why are the scholars trained to do this historians and not philosophers, according to Kuhn? Because philosophers had imbibed a general claim that theories are simply sets of laws—empirical generalizations and higher-level covering laws for the same. But this illuminated for Kuhn neither the identity nor the function of theory in science. He emphasized throughout a form of theory holism and, in this essay, a sort of theory universalism. That is, not only were the concepts of theory related each to the other so you could not atomize the theory into distinct semantic or epistemic parts that could be improved piece-meal, but theory, whenever it appeared, appeared in a totalizing form—all of the subject matter of science was constrained by theory. Their underestimation of the role of theory in science and their bland characterization of theory as sets of laws or lawlike sentences disqualified philosophers from the project of being

---

[2] Or, again, history is no more a set of facts for Kuhn than a theory is for him a set of laws. His main concern is not with the question of which facts philosophy of science is beholden to.

able properly to reconstruct the processes of theory development and evaluation in science.

4.3   A Puzzle

By now, the reader might be in some difficulty. There seem to be several claims on the table now that stand in some tension. First, we can recognize different interests and concerns among philosophers and historians of science. Second, among the concerns of the philosophers are the universal and the normative in science. Third, theory always and everywhere operates in science. Fourth, delineating the processes of theory development and evaluation (of which at least the latter is surely normative) is more securely a project for historians than for philosophers (since philosophers import a wrong and simplified notion of theory from their own concerns). Historians thus seem to be better able to do the normative work that only the philosophers actually care about. This is a reduction if not to the absurd, then at least to the unfortunate and dysfunctional.

This puzzle is not, so far, terribly deep or troublesome. It simply requires a bit more accuracy of expression to see that it is more apparent than real. But it does lead to one observation and one project, both of which are aspects of, if not the troubled marriage, then the troubled history of the connections between history and philosophy of science. However troubled this history, I do think the proposed project, suitably historically informed, is interesting.

First, let us clarify the situation and make some distinctions. It is illusory that Kuhn, in saying what he has said (and I have reported), is somehow assigning normative interests to philosophers but the corresponding normative tasks to historians. Historians are not themselves evaluating scientific claims when they reconstruct the processes of development and evaluation of theories: they are not doing normative history; they are doing the history of normative practices. Scientists (and presumably at least in some places and times, philosophers) evaluate and develop theories. Historians reconstruct the histories of such practices—they are able, then, to exhibit the normative structures of science in its various places, times, objects, etc. This does mean that normative practices have to be recognizable as normative—theory evaluation must be recognizable as evaluation. (It might not, however, be separable even in principle from theory discovery, development, or articulation. That is why Kuhn rejects the discovery/justification distinction.) Philosophical concern with normativity then looks somewhat peculiar but not, in principle, impossible—it is perhaps an attempt to evaluate the theories of one place and time relative to standards of another, or it might be the attempt to evaluate science by the standards of philosophy. One can do this but, in so doing, one has ceased to be interested in understanding what actually happened in the history of science.

With the articulation of a difference between undertaking a normative task and undertaking a history of normative tasks, the question of the normative is relatively

straight-forwardly dealt with. The question of the universal is rather more complicated. As the title of Kuhn's book suggests, Kuhn really does seem to offer a universal
theory of science: there are structures that are always and everywhere articulated in
science. Science always involves theory, traditions of puzzle-solving, articulation
(once sciences reach a certain level of development or maturity) of puzzle-solving
paradigms, anomalies, revolutions. The historical details that go into filling out these
formal features will be specific and context-bound; the structures themselves are
universal. Thus, it cannot be true that philosophers do, but historians do not, concern
themselves with the universal in science. There is, for Kuhn, a universal form of
historical narration in virtue of which an episode is recognizable as an episode in the
history of science.

Presumably, then, in characterizing his philosophy students as being transhistor-
ical universalists (not his term), Kuhn wishes to characterize them as being more
universalist in their tendencies than he himself and his historical students were. That
does not seem implausible. Often philosophers write as if there are universal standards of rationality that apply everywhere and everywhen. If you have such a view
you don't just need to tell a structurally compliant story of, say, Newton's or Darwin's
science but also you will seek to evaluate their achievements relative to a universal
standard, rather than relative to the historically available standards (which standards
cannot even be articulated independently of articulating the science whose standards
they constitute). This sort of philosophical universalism sounds to historians ever so
much like historical whiggism precisely because it also (almost) invariably comes
with the further commitment that says that the universal standards are now more
well-understood and -articulated than ever before.

4.4   A More Interesting Puzzle and a Project for HPS Fifty-Odd Years On

The specifics of our original puzzle have turned out not to be that puzzling. But
one lesson embedded in how we have made that puzzle less puzzling is that Kuhn's
language for talking about history and philosophy is not always quite well-chosen.
History is not an entirely anti-universalist project for Kuhn; nor does it ignore and,
thus, leave to the philosophers everything in science that might properly be called
"normative." This lack of discursive fit infected the early discussions of the relations
between history and philosophy of science more generally. Thus, for example, Ron
Giere in the very essay that (it seems) bequeathed to us the marriage metaphor offered
up this odd characterization of the work of Kuhn (and Feyerabend):

> Turning to the problem of validation, I think a majority of philosophers of science, contrary
> to Kuhn, Feyerabend, and others, agree that there is such a thing as empirical validation.
> That is, there is such a thing as non-deductive reasoning. (Giere 1973, p. 294)

Those were heady times, but it is rather astonishing when reading Kuhn's early work
to think that anyone would have read Kuhn as arguing that scientific theories did

not undergo something that might reasonably be called 'empirical validation.' It is far more accurate to say that Kuhn would be unimpressed by Giere's collapse of the question of empirical validation to the question of 'non-deductive reasoning.' This sets aside all sorts of interesting aspects to the question of the empirical success of a theory—such as the lessons about commitment to the theory and of a scientist proving her own mettle through her measurement results outlined in, for example, his "The Function of Measurement in the Modern Physical Sciences" (Kuhn 1961/1977a). We have, in any case, seen that the history of science includes an interest in the history of theory evaluation for Kuhn—'inductive logic' is not Kuhn's own project but presumably the interrelations of, for example, nineteenth-century British Science and nineteenth-century British inductive logics would be an appropriate topic of historical research for Kuhn.

In the volume Giere was reviewing, just to take up another example, Herbert Feigl (Feigl 1970) framed his account of how history and philosophy of science might come together productively by invoking the justification/discovery distinction. The projects come together by jointly offering a more complete account of science than either on its own could offer, given their different subject matters. Feigl, thus, endorsed just the distinction that Kuhn argued against. There was not, at the time of the early post-Kuhnian attempt to theorize the relations of history of science and philosophy of science, an agreed upon language within which proponents of various views might discuss the conceptual issues. Of course, in a world in which there are disagreements, finding a common language is not easy—it is notable just how impoverished the language in which the conceptual relations between history and philosophy of science were formulated was during this formative period (and how it is really no better now). This is the puzzle.

The project I'd like to suggest is not the articulation of a 'neutral metalanguage' that will allow such conceptual conversations to take place more fruitfully. That is a pipe dream. Rather, I'd like to suggest that Kuhn actually did, in some of his formulations, offer some very helpful ways of speaking that might allow us more productively to think about how history and philosophy of science might get on with one another now (and with other component projects of science and technology studies). I have in mind Kuhn's emphasis on the *writing* of history and of philosophy and, thus, on the products of historical scholarship and philosophical analysis.[3] Here is Kuhn once again:

> The final product of most historical research is a narrative, a story, about particulars of the past. In part it is a description of what occurred (philosophers and scientists often say, a *mere* description). Its success, however, depends not only on accuracy, but also on structure. The historical narrative must render plausible and comprehensible the events it describes. . . . The philosopher, on the other hand, aims principally at explicit generalizations and at those with universal scope. He is no teller of stories, true or false. His goal is to discover and state what is true at all times and places rather than to impart understanding of what occurred at a particular time and place. (Kuhn 1977b, p. 5)

---

[3] Cf. J. Schikore (2009).

What is helpful in such passages is Kuhn's insistence on talking about the methods and the results of historical and philosophical work—what do we actually produce in our work and how do we do so.

Another thing that is clear in passages such as this is that Kuhn's account of the goals of history are a great deal more understandable and, seemingly, more capable of being successfully discharged than are the goals and tasks he ascribes to philosophy. There is something very odd about this vision of philosophers, however, as aiming for eternal verities. There have been philosophers who have thought this way but they are not notably well-represented among, for example, logical empiricist philosophers of science. A more helpful response to such passages in Kuhn than the early Giere's or McMullin's discourses on the relations between history and philosophy of science would be a more careful articulation of the processes and products of philosophical work.

For what it is worth, as someone who has written the odd philosophical paper, I find that Kuhn does not here come close to describing what it feels like to write an essay in the effort to achieve philosophical understanding. Indeed, the general dichotomies that proliferate in the early discourses on history and philosophy of science—the universal and the particular, the timeless and the timebound, the normative and the descriptive—seem more to obscure than to illuminate the processes of philosophical thinking. For me, a more accurate articulation would be in the language of the conceptual maps of various intellectual regions—to get clear on, say, the epistemological status of testimony is to understand the relations of testimony to other related notions. Often enough, a philosopher wishes to state a coherent position on, say, the status of reasons, without committing to that position. Rather than implicitly holding a universalist or eternalist view of such conceptual maps, in my experience, philosophers are indifferent to the questions of the universal and the particular, the timeless and the time-bound. Moreover, while the methods perhaps suggest that philosophical understanding strives for the universal, contrary to the suggestion in Kuhn, philosophy rarely expresses itself in the language of universal generalizations. The living practices of philosophers are seriously under- and mis-described in such remarks.

Moreover, whatever the goal of philosophical understanding, it is implausible to think that philosophical problems are not as temporally-conditioned as problems in any other area of human life. Certainly, for example, the philosophical means for achieving clarity (and the clarity achievable by such means) about induction differ widely in Hume, Mill, and Carnap. The mind/body problem, to take another well-worn example, arose as a genuine problem in philosophy only with the modern era and the rise of the "representational theory of mind"; ancient philosophers could philosophize perfectly well without taking a side on the mind/body problem. Only recently has consciousness become not a resource for solving problems but rather a problem for philosophy.[4]

---

[4] Indeed, consciousness has become a series of ever harder problems, needing an ever more rugged group of philosophers of mind to handle [it?].

So, the project I am imagining is one in which historians and philosophers of science do not fret so much about the nature of their relationship as they try better to articulate their own explanatory practices. Since, on this score, philosophical understanding is less well articulated than is historical understanding, philosophers of science in particular would do well to explain more what they are trying to do in advancing a specifically philosophical understanding of science, what resources they bring to bear in drawing out that understanding, what issues they take up and what issues they set aside. One specific question suggested by Kuhn's *Structure* is this: what is the relation between the scientist's understanding of science and the philosopher's understanding of science? If philosophy of science is not where the scientist's understanding of science is codified and made explicit, what is it? What is it for? Where, if anywhere, in either history or philosophy of science is a critique of science located?

4.5 Coda: A Place for History in Philosophical Understanding

In his various attempts to work out the conceptual structure of theories, Kuhn returned again and again to holistic metaphors and functionalist language, trying to express how the same principles provided a vocabulary with which to talk meaningfully at all about nature and, at the same time, expressed a set of specific claims that ruled, as it were, a priori with respect to nature so semantically circumscribed. Facts about Kuhn's own biography and an increased knowledge of neo-Kantian projects in philosophy of science suggest something both resonantly and problematically Kantian in such sensibilities. Kuhn's remarks on the differences between the sensibilities of historians and philosophers of science are tempered, for example, by his own repeated insistence that the sensibilities of historians of science were themselves importantly forged or fostered by the work of neo-Kantian philosophers, such as Ernst Cassirer and Emile Meyerson. If Kuhn had been less time-bound in his own understanding of philosophy, he might have seen that specifically the question of what made a narrative of the development of scientific knowledge a narrative of progress, of science, and of knowledge was itself a prominent question in philosophy for well over a hundred years after Kant, that, in a sense, professionalized history of science developed out of a now displaced project in philosophy of science.[5]

Within the American context, history and philosophy of science has reached its second half-century of institutionalized existence; science and technology studies is only a couple of decades younger. Many of the arguments within those fields have been about either the object of mutual interest (science) or about the very nature of interdisciplinary work. Some of those arguments have been productive,

---

[5] Not wholly displaced, however. There are both explicitly neo-Kantian philosophies of science (Friedman 2001; Domski and Dickson 2010) in the contemporary scene and a burgeoning literature in the neo-Kantian philosophers such as Ernst Cassirer (e.g. Friedman 1999; Mormann 2008; Heis 2014) that Kuhn himself points to as antecedents to professional history of science.

some less so. It is the argument of this essay that Kuhn pointed the way to a more productive avenue of discussion in history and philosophy of science and science and technology studies, even if he did not himself always take his own advice: if we become clearer as disciplines on what our own practices of achieving understanding are and what questions those practices do or do not, might or might not, answer, had more interest in our practices of producing and reading texts, we'd have a better mutual understanding. It is an argument for a form of productive reflexivity that is aimed at communication—not an effort to disappear up our own navels but an effort to articulate what we are actually trying to do and to take seriously our own failures of expression when they become evident.

References

Burian, R. M. 1977. More than a marriage of convenience: On the inextricability of history and philosophy of science. *Philosophy of Science* 44:1–42.

Daston, L. 2009. Science studies and the history of science. *Critical Inquiry* 35:798–813.

Domski, M., and M. Dickson, eds. 2010. *Discourse on a new method: Viagra for the marriage of history and philosophy of science*. Chicago: The Open Court.

Feigl, H. 1970. Beyond peaceful coexistence. In *Historical and philosophical perspectives on science*, ed. R. H. Stuewer, 3–11. Minneapolis: University of Minnesota Press.

Friedman, M. 1999. *A parting of the ways: Carnap, Cassirer, Heidegger*. Chicago: The Open Court.

Friedman, M. 2001. *Dynamics of reason*. Stanford: CSLI.

Giere, R. N. 1973. History and philosophy of science: Intimate relationship or marriage of convenience? *British Journal for the Philosophy of Science* 24:282–297.

Giere, Ronald N. 2011. *History and philosophy of science: Thirty-five years later*. In *Integrating history and philosophy of science*, ed. S. Mauskopf and T. Schmaltz, 59–65. New York: Springer, (BSPS 263).

Heis, Jeremy. 2014. Ernst Cassirer's *Substanzbegriff und Funktionsbegriff*. *HOPOS* 4: 241–270.

Hesse M. B. 1961. *Forces and fields: The concept of action at a distance in the history of physics*. London: Thomas Nelson and Sons.

Hesse M. B. 1963. Review of *The structure of scientific revolutions* by Thomas S. Kuhn. *Isis; an international review devoted to the history of science and its cultural influences* 544:286–287.

Kuhn, T. S. 1962. Untitled review of *Forces and fields* by Mary B. Hesse. *American Scientist* 50:442A–443A.

Kuhn, T. S. 1961/1977a. The function of measurement in modern physical science. In *The essential tension*, ed. T. S. Kuhn, 178–224. Chicago: University of Chicago Press.

Kuhn, T. S. 1977b. The relations between the history and the philosophy of science. In *The essential tension*, ed. T. S. Kuhn, 3–20. Chicago: University of Chicago Press.

Kuhn, T. S. 2000. *The road since structure*. Chicago: University of Chicago Press.

Kuhn, T. S. 1962/2012. *The structure of scientific revolutions*. Chicago: University of Chicago Press.

McMullin, E. 1976. History and philosophy of science: A marriage of convenience? In *PSA 1974*, ed. R. S. Cohen, et al. 585–601. Dordrecht: Reidel.

Mormann, T. 2008. Idealization in Cassirer's philosophy of mathematics. *Philosophia Mathematica* 16:151–181.

Schickore, J. 2009. Studying justificatory practice: An attempt to integrate the history and philosophy of science. *International studies in the Philosophy of Science* 23 (1): 85–107.

Chapter 5
Reconsidering the Carnap-Kuhn Connection

Jonathan Y. Tsou

5.1 Introduction

Rudolf Carnap (1891–1970) and Thomas Kuhn (1922–1996) are undoubtedly two of the most influential twentieth-century philosophers of science. According to the 'received view' on the Carnap-Kuhn relationship, Carnap's and Kuhn's views represent diametrically opposed approaches to philosophy of science and Kuhn's ([1962] 1996) *Structure of Scientific Revolutions* (henceforth, *Structure*) is one of the main philosophical works—along with W.V.O. Quine's ([1951] 1980) "Two Dogmas of Empiricism"—that (rightfully) contributed to the demise of logical empiricism in the 1960s and 1970s. While the received view has been commonplace among post-positivist philosophers of science (e.g., see Suppe [1974] 1977; Giere 1988, Chap. 2; McGuire 1992), this narrative about the history of philosophy of science has been increasingly called into question in recent decades.

Some historians of philosophy of science (Reisch 1991; Earman 1993; Irzik and Grünberg 1995; Friedman 2001, 2003; Irzik 2002, 2003; Richardson 2007; Gattei 2008, Chap. 5; Uebel 2011) have argued that the received view on Carnap and Kuhn is mistaken, suggesting that there is a close affinity between their philosophical views. The basis for this revised understanding stems from some fundamental similarities between the philosophical systems of Carnap and Kuhn, especially on issues concerning incommensurability, theory-choice, and the nature of scientific revolutions. The upshot of this revisionist picture is that the "two styles of doing philosophy of science epitomized by Carnap and Kuhn should be seen as complementary rather than mutually exclusive" (Irzik and Grünberg 1995, pp. 304–305). Furthermore, some revisionists have drawn the more radical conclusion that this revised understanding of the relationship between Carnap and Kuhn "undermines the widely held belief that post-positivist philosophy of science represents a revolutionary departure

J. Y. Tsou (✉)
Department of Philosophy and Religious Studies, Iowa State University, Ames, IA, USA
e-mail: jtsou@iastate.edu

© Springer International Publishing Switzerland 2015   51
W. J. Devlin, A. Bokulich (eds.), *Kuhn's Structure of Scientific Revolutions—50 Years On*,
Boston Studies in the Philosophy and History of Science 311,
DOI 10.1007/978-3-319-13383-6_5

from its arch-rival positivism, at least in the context of Carnap's and Kuhn's works" (Irzik and Grünberg 1995, p. 304).

In this chapter, I argue against the revisionist conclusion that Carnap's and Kuhn's philosophical views are closely aligned; moreover, I reject the revisionist idea that Kuhn's philosophical views do not represent a revolutionary departure from Carnap's.[1] While there are undoubtedly similarities between Carnap's and Kuhn's philosophical systems, I argue that a consideration of their broader philosophical projects renders these similarities superficial in comparison to their fundamental differences. On a general level, revisionist analyses fail to sufficiently acknowledge that Carnap's linguistic frameworks are logical reconstructions *intended to clarify answerable (i.e., meaningful) and unanswerable (i.e., meaningless) questions*, while Kuhn's theory of scientific revolutions is *motivated to provide a naturalistic description of scientific change*. This difference reflects two vastly different styles of doing philosophy of science (viz., logical analysis versus historical analysis). On a more specific level, I argue that Carnap's stance on incommensurability is far less robust than Kuhn's, Carnap holds a more instrumentalist position on theory-choice than Kuhn, and Carnap's analysis of revolutions is antithetical to Kuhn's. From this perspective, I suggest that the methodologies of Carnap and Kuhn are correctly regarded as two contrasting philosophical styles that mark a significant division between positivist and post-positivist philosophy of science.

5.2   The Revisionist View

The basis for the revisionist view stems from parallels in Kuhn's theory of scientific revolutions and Carnap's philosophy of linguistic frameworks (Friedman 2001, pp. 41–43; 2003, pp. 20–22; Richardson 2007, pp. 354–356). Kuhn's claim that the solution to scientific puzzles is provided by the tacit rules of a paradigm is similar to Carnap's ([1950] 1956) claim that answers to meaningful (internal) questions are provided by the rules of a linguistic framework. Moreover, just as Kuhn maintains that there are no clearly defined (algorithmic) rules for choosing among competing paradigms, Carnap holds that there is no cognitively significant (i.e., meaningful) way of choosing among alternative linguistic frameworks. On this issue, both Kuhn and Carnap maintain that these decisions must be made on non-epistemic grounds. For revisionists, these similarities suggest a significant point of agreement between

---

[1] At the outset, it is important to note that there is variability among how strongly (and how qualified) the revisionist thesis is advanced by various authors. Moreover, different revisionist analyses have been forwarded for various purposes, e.g., Friedman's (2001, 2003) analysis is motivated to demonstrate a shared neo-Kantian heritage inherited by Carnap and Kuhn (see DiSalle 2002; Richardson 2002; Tsou 2003; Lange 2004). The main target of the argument in this paper is Gürol Irzik and Teo Grünberg's (1995) influential article, "Carnap and Kuhn: Arch Enemies or Close Allies?", which offers one of the strongest expressions of the revisionist view. While my argument focuses on Irzik and Grunberg's (1995) article, it is more broadly applicable to other revisionist analyses that, either explicitly or implicitly, follow a similar line of reasoning.

Carnap and Kuhn grounded in "a pragmatically oriented semantic conventionalist picture of science" (Irzik and Grünberg 1995, p. 285).

Revisionists contend that both Carnap and Kuhn endorse a version of the incommensurability thesis (Irzik and Grünberg 1995; Irzik 2002; Richardson 2007, pp. 356–357). Incommensurability is central to Kuhn's ([1962] 1996, Chaps. 9–10) idea that competing paradigms are *incompatible* to the extent that proponents of competing paradigms cannot communicate with one another since their theoretical and epistemic commitments preclude them from *comprehending* alternative views. In *Structure* (Kuhn ([1962] 1996), incommensurability variably refers to the incompatibility of *problems and standards* (p. 103, 148–149), *meaning* (pp. 101–103, 149), and *perception* (p. 112, 150). In post-*Structure* writings, Kuhn (2000) offers a more precisely defined thesis of 'local incommensurability' in terms of untranslatability: "The claim that two theories are incommensurable is... the claim that there is no language, neutral or otherwise, into which both theories, conceived as sentences, can be translated without residue or loss" (p. 36). To support the idea that Carnap endorses a similar thesis, Irzik and Grünberg (1995, pp. 291–295) point to Carnap's ([1936] 1949) claim that competing linguistic frameworks are sometimes *untranslatable*:

> In translating one language into another the factual content of an empirical statement cannot always be preserved unchanged. Such changes are inevitable if the structures of the two languages differ in essential points.... [W]hile many statements of modern physics are completely translatable into statements of classical physics, this is not so... when the statement... contains concepts (like, e.g., 'wave-function' or 'quantization') which simply do not occur in classical physics... [T]hese concepts cannot be... included since they presuppose a different form of language. (p. 126)

Irzik and Grünberg (1995) contend that this 'semantic untranslatability' thesis is essentially the same as Kuhn's local meaning incommensurability thesis and Carnap's endorsement of this thesis follows from his commitment to a semantic holism (i.e., that the theoretical postulates of a linguistic framework determine the meaning of theoretical terms in *L*) similar to Kuhn's holism (pp. 291–293).

Other revisionists emphasize that both Carnap and Kuhn maintain that choosing between competing scientific theories is a non-epistemic and pragmatic matter (Friedman 2001, pp. 41–43; 2003, pp. 19–21).[2] In Carnap's philosophy, this stance is explicit in the understanding of external questions as *practical proposals* to adopt a particular linguistic framework. Carnap ([1950] 1956) writes:

> [T]he introduction of [a new linguistic framework] does not need any theoretical justification because it does not imply any assertion of reality... [W]e have to face at this point an important question; but it is a practical, not a theoretical question... of whether or not to accept the new linguistic forms. The acceptance cannot be judged as true or false because it

---

[2] Gürol Irzik (2003) has argued—correctly, in my judgment—against this specific claim. In particular, Irzik opposes "relativist" interpretations of Carnap and Kuhn (e.g., see Friedman 1998, 2001), suggesting that both Carnap and Kuhn hold nuanced views on scientific rationality that are not accurately described as "relativist" (cf. Axtell 1993; Irzik 2003, pp. 331–335).

is not an assertion. It can only be judged as being more or less expedient, fruitful, conducive to the aim for which the language is intended. (p. 214)

For Carnap, choosing a linguistic framework only implies a commitment to a particular way of speaking. Since linguistic frameworks can be employed for different purposes, Carnap believes that they should be evaluated as instruments for various ends, rather than by their 'correctness.' In the spirit of the 'principle of tolerance,' Carnap recommends a permissive and pluralistic attitude towards different linguistic forms (Carnap [1934] 1937, § 17; [1950] 1956, p. 221; Jeffrey 1994). Kuhn ([1962] 1996, pp. 94–110, 198–207; 1977; 2000, Chap. 9) adopts a similar stance on the non-epistemic nature of theory-choice insofar as he argues that choosing between competing paradigms is a process that cannot be settled in terms of 'correctness.' Kuhn (1977) emphasizes that, in comparing the relative merits of competing paradigms, scientists typically appeal to a set of fixed values (e.g., empirical adequacy, consistency, explanatory scope, simplicity); however, when applying these values, proponents of different paradigms will interpret and place different weights on these values. Hence, there is no objective (i.e., shared) set of values that can be appealed to in theory choice, which necessarily involves appeals to subjective factors. Insofar as Kuhn holds that there is no truly objective (or intersubjective) basis for paradigm choice, he endorses the Carnapian view that theory-choice is ultimately non-epistemic and pragmatic.

Revisionists also suggest that Kuhn and Carnap share a similar view of scientific revolutions (Reisch 1991; Irzik and Grünberg 1995; Friedman 2001, p. 22, 41–42; Irzik 2002). Kuhn ([1962] 1996, chs. 9–10) famously rejects the view that scientific change is continuous and cumulative. In his model, scientific change proceeds through repeated cycles of normal science and revolutionary science. Whereas normal science is a cumulative period of puzzle-solving, revolutionary science is characterized by an older paradigm being *replaced* (in whole or in part) by an incommensurable new one. Hence, Kuhnian revolutions are neither rule-governed nor cumulative, which opposes the putative view that scientific change is progressive and cumulative. Revisionists suggest that Carnap endorses a similar view of revolutions. In discussing how scientists respond to anomaly (a Quinean 'recalcitrant experience'), Carnap (1963) summarizes his view of revolutions as follows: "[A] change in the language. . . constitutes a radical alteration, sometimes a revolution, and it occurs only at certain historically decisive points in the development of science. . . . A change of [this] kind constitutes, strictly speaking, a transition from a language L_n to a new language L_{n+1}" (921). Given Carnap's views on semantic untranslatability and the pragmatic nature of theory choice, the transition from one linguistic framework to another is a process governed by pragmatic factors, and hence, discontinuous and non-cumulative.

5.3   A Problem with the Revisionist View

While the revisionist view reveals some interesting similarities between Carnap's and Kuhn's philosophical views, when Carnap's and Kuhn's views are examined in the context of their broader philosophical projects, these similarities turn out to be superficial rather than substantial. Revisionist analyses articulate their arguments by framing Carnap's views in Kuhnian terminology, such as 'paradigms,' 'incommensurability,' and 'scientific revolutions.' Moreover, they suggest that Carnap's linguistic frameworks can be understood as an analogue (or formal complement) to Kuhn's paradigms (in the sense of 'disciplinary matrices'). This exegetical perspective, however, obscures the fundamentally disparate nature of the broader philosophical projects of Carnap and Kuhn. In particular, it fails to sufficiently acknowledge that Carnap's linguistic frameworks are *artificial languages that scientific philosophers construct for purposes of logical analysis*, while Kuhn's paradigms are conceived of naturalistically, *as a constellation of commitments (i.e., symbolic generalizations, metaphysical commitments, values, exemplars) shared by a community of scientists.* Framing Carnap's linguistic frameworks as analogues to Kuhn's paradigms brings Carnap's philosophical project closer to Kuhn's agenda of providing an *accurate historical description of actual scientific practices and theories*; however, it does so at the expense of obscuring the fundamental nature and aims of Carnap's philosophy. In what follows, I explicate the nature of Carnap's logic of science program to motivate an argument that the fundamental differences between Carnap's and Kuhn's broader philosophical projects render the similarities that revisionists highlight superficial.

The proper context for understanding Carnap's philosophy of linguistic frameworks is Carnap's "logic of science" (*Wissenschaftslogik*), which is Carnap's proffered replacement for 'epistemology' or 'philosophy' more generally (see Richardson 1998, Chap. 9). In *Logical Syntax of Language*, Carnap ([1934] 1937) presents the logic of science as a scientific philosophy characterized by logical analysis:

> That part of the work of philosophers which may be held to be scientific. . . consists of logical analysis. The aim of logic of science is to provide a system of concepts, a language, by the help of which the results of logical analysis will be exactly formulable. *Philosophy is to be replaced by the logic of science*—that is to say, by the logical analysis of the concepts and sentences of the sciences, for *the logic of science is nothing other than the logical syntax of language.* (p. xiii, emphasis in original)

According to Carnap, the task of scientific philosophers is to logically analyze scientific concepts and sentences. This methodological prescription is motivated to ensure meaningful discourse about science. Carnap believes that modern logic provides the necessary tools to transform (or translate) formerly metaphysical problems into meaningful problems. In "Unity of Science," Carnap ([1931] 1934) rejects the traditional fields of philosophy (i.e., metaphysics, epistemology, and ethics) and describes the problems of scientific philosophy as follows:

> [O]ur own field of investigation is that of *Logic*. Here are to be found problems of. . . the *Logic of Science*, i.e., *the logical analysis of the terms, statements, theories, proper to the various department[s] of science.* Logical Analysis of Physics, for example, introduces the

problems of Causality, of Induction, of Probability, the problem of Determinism... [as] question[s] concerning the logical structure of the systems of physical laws, in divorce from all metaphysical questions.... Logical Analysis of Biology, again, involves the problems of Vitalism, to take one example... in a form free from Metaphysics, viz. as a question of the logical relations between biological and physical terms and laws.... In Psychology, Logical Analysis involves, among others the so called problem of the 'relation between Body and Mind'... concerned... with the logical relations between the terms or laws of Psychology and Physics respectively.... *In all empirical sciences, finally, Logical Analysis involves the problem of verification... as a question concerning the logical inferential relations between statements in general and so called protocol or observation statements.* (pp. 24–25, emphasis added)

As indicated in the last sentence of this passage, Carnap maintains that the logic of science is especially concerned with the problem of "verification" or "confirmation," which is an aspect of Carnap's philosophy that is consistently neglected by revisionists.[3] For the purposes of this paper, I want to highlight the deflationary nature of logic of science and indicate how it relates to the problem of meaningfulness. By reformulating and reconstructing scientific theories into purely logical (i.e., syntactic and semantic) systems or linguistic frameworks, Carnap believes that meaningful discourse about science can be ensured *by clarifying the empirical basis of scientific theories* (see Dempoulos 2003, 2007). In the case of sciences such as physics and biology, this amounts to translating metaphysical problems into empirically ascertainable ones.

One of the main tasks of the logic of science is to develop an exact and objective method (the method of 'logical syntax') for discussing scientific propositions. For Carnap ([1934] 1937), the "important thing is *to develop an exact method for the construction of... sentences about sentences*" (p. xiii, emphasis added). The logical syntax of a language is simply the "formal theory of linguistic forms of that language—the systematic statement of formal rules which govern it together with the development of the consequences which follow from these rules" (§ 1). By reconstructing a language into its syntax, Carnap believes that one can specify the rules of a language. Carnap proposes to construct sentences about sentences by constructing two languages: (1) the object-language, which is the language that is the object of investigation (e.g., a scientific theory), and (2) the syntax-language (or meta-language), which is the language used to speak *about* the object-language. Carnap (§§ 78–81) believes that confusion occurs when philosophers speak within an object-language (the so-called *material mode of speech*) without recognizing that these assertions are made within or relative to an object-language. For Carnap, the meta-language (the *formal mode of speech*)—the perspective that reconstructs sentences and concepts of the object language syntactically—is the proper (and metaphysically neutral) philosophical perspective for evaluating these sentences, *not as assertions*, but as *proposals* to use a linguistic framework.

---

[3] Carnap's preoccupation with these issues is most clearly represented in his various (and increasingly deflationary) attempts at articulating an empiricist criterion of meaningfulness (see Carnap [1931] 1959, 1936, 1937, 1956; Hempel 1965).

Given the nature and aims of the logic of science, it is important to see that revisionists employ a self-serving exegetical strategy when they frame Carnap's philosophy of linguistic frameworks in Kuhnian terms. In particular, it is mistaken to regard Carnapian linguistic frameworks as straightforward analogues (or even formal complements) to Kuhnian paradigms. This interpretation is suggested by phrases like "every scientific theory is embedded within a linguistic framework" (Irzik 2002, p. 607); "a shift from one linguistic framework to another is a revolution" (Irzik and Grünberg 1995, p. 295); or "[p]aradigms, like linguistic frameworks, constitute the conditions of scientific knowledge—scientific knowledge-making only unproblematically occurs when a paradigm is in hand" (Richardson 2007, p. 336). These characterizations suggest that Carnapian linguistic frameworks can be understood as akin to Kuhnian paradigms, i.e., as *a historically-situated set of commitments and assumptions that function to bind scientific communities* (cf. Irzik and Grünberg 1995, p. 286). This understanding, however, inverts Carnap's philosophy of linguistic frameworks. Carnapian frameworks are not (temporally or logically) prior to scientific theories, but theories are prior to linguistic frameworks insofar as the latter are *logical reconstructions of scientific theories*, which are formulated to clarify the meaningful basis of theories. Conversely, *from a naturalistic perspective*, Kuhn ([1962] 1996, Ch. 5) regards paradigms as (temporally and logically) prior to scientific rules and theories.[4] This subtle difference highlights a significant contrast between Kuhnian paradigms (i.e., a cluster of shared commitments that are necessary for normal science) and Carnapian linguistic frameworks (i.e., formal reconstructions of scientific theories). While revisionists are well aware of the artificial nature of linguistic frameworks (Irzik and Grünberg 1995, p. 288; Friedman 1999, Ch. 9; Richardson 1998, Ch. 9), when they argue for similarities between Carnap and Kuhn, they obscure the nature of Carnap's linguistic frameworks by presenting them in a Kuhnian light.[5]

---

[4] In Chap. 5 of *Structure* ("The Priority of Paradigms"), Kuhn argues that it is paradigms (rather than explicit rules) that determine the nature of normal science. Kuhn suggests that paradigms are prior to rules in a temporal sense (i.e., paradigms will suggest certain rules, but not in a determinate way), but also in terms of importance (i.e., paradigms are more important than the rules that are abstracted from the paradigm for binding a community of scientists during normal science). On the basis of these considerations, Kuhn suggests that philosophers of science ought to focus their attention on paradigms (i.e., exemplars), as a unit of analysis, rather than explicit rules. It is important to notice that Kuhn's methodological prescription is opposed to Carnap's attempt to reduce scientific theories to a set of explicit rules (e.g., syntax). Moreover, the tacit rules discussed by Kuhn are not the same kinds of rules at the core of Carnap's linguistic frameworks (Pincock 2012, pp. 127–128).
[5] From a somewhat different perspective, Peter Galison (1995) suggests that Kuhnian paradigms and Carnapian linguistic frameworks are similar insofar as they represent science in terms of "island empires," i.e., isolated and relatively stable assemblages of experimental and theoretical procedures and results. Galison opposes this island empire picture of science because it conceals the fragmented and heterogeneous nature of science.

5.4   Logical Analysis vs. Historical Analysis

Differences in Carnap's and Kuhn's methodological assumptions reflect two radically contrasting styles of doing philosophy of science. In the following section, I dub these two styles *logical analysis* and *historical analysis*, and I articulate the assumptions of these fundamentally disparate ways of doing philosophy of science. In the context of the revisionist argument, this shows that there are good reasons for regarding Carnap's and Kuhn's views as standing in an antagonistic, rather than complementary, relationship.

The most fundamental difference between Carnap's and Kuhn's styles of doing philosophy of science is their methodological approaches to analyzing scientific theories. For Carnap, logical analysis assumes that scientific theories should be analyzed only after they have been reconstructed into artificial linguistic frameworks, which scientific philosophers can investigate in the formal mode of speech. Kuhn's approach, by contrast, assumes that theories should be analyzed after they have been historically reconstructed as paradigms or lexicons. The key difference in Kuhn's approach is that paradigms are treated and analyzed, not as artificial languages, but as *naturalistic entities*, i.e., as *accurate descriptions of scientific theories*.[6] Hence, Kuhnian philosophy of science is concerned with accurately reconstructing scientific theories and practices with the aid of *a posteriori* sciences such as history and psychology (see Giere 1985; Bird 2002; 2004; Preston 2004). This naturalistic aspect of Kuhn's approach is entirely absent in Carnap's. Whereas accurate historical reconstruction, for Kuhn, is crucial for proper philosophical analysis, Carnap is only concerned with accurate reconstruction to the extent that it allows him to distinguish theories into their observational and theoretical parts, which will clarify the sense in which theories are *cognitively significant*.

Carnap and Kuhn also adopt contrasting stances on the context of discovery and context of justification distinction (Pinto de Oliveira 2007; cf. Uebel 2011). Whereas Carnap assumes a sharp distinction (given that logic of science occurs exclusively in the context of justification), Kuhn believes that there is no sharp distinction and that issues concerning the justification of scientific theories cannot be analyzed in isolation from issues regarding the discovery of those theories (Kuhn [1962] 1996, pp. 8–9,

---

[6] As a qualification, Kuhn's naturalistic approach was most marked in *Structure*, and in his post-*Structure* writings. Kuhn took a 'linguistic turn' wherein his work became more traditionally philosophical and relied less heavily on the history of science (see Irzik and Grünberg 1998; Bird 2000, 2002, 2004; Kindi 2005; Mladenović 2007; Gattei 2008). In offering a qualified defense of the received view on the Carnap-Kuhn relationship, my claim is that Kuhn's early philosophical views—as exemplified in *Structure*—are significantly different than the style of philosophy of science championed by Carnap. While Irzik and Grünberg (1995) focus on Kuhn's later work in advancing their argument that Carnap and Kuhn are 'close allies,' I focus on Kuhn's early views because: (1) Kuhn's *Structure* was much more influential and widely read by philosophers of science than his later works, and (2) Kuhn's *Structure* is the most relevant work for the received view on the Carnap-Kuhn relationship that maintains that Kuhn contributed to the demise of logical empiricism by offering a revolutionary approach to philosophy of science (discussed in Sect. 5.5 of this chapter).

207–208). Carnap's ideal of rationally reconstructing scientific theories into linguistic frameworks assumes that the philosopher can investigate questions concerning the justification of scientific theories via logical analysis by clarifying which parts of the theories are empirically ascertainable. Issues concerning the justification of theories for Kuhn are more complex.[7] In response to questions concerning whether his theory is descriptive or prescriptive (e.g., see Feyerabend 1970), Kuhn ([1962] 1996, pp. 207–208; 2000, p. 130) argued that descriptive generalizations from the history of science can sometimes serve as *evidence* for philosophical prescriptions. This aspect of Kuhn's philosophy highlights the way in which his approach rejects the discovery/justification distinction and takes actual scientific practices seriously. Whereas Carnapian logical analysis takes reconstructed artificial linguistic frameworks (removed from the context of discovery) as the proper unit of philosophical analysis, Kuhnian historical analysis takes actual scientific theories and practices to be the proper unit of analysis (from which prescriptive claims can subsequently be inferred). Whereas Carnap's philosophical system prescribes certain scientific standards *a priori* (e.g., theories should be non-metaphysical and empirically meaningful), Kuhn's philosophy takes a more *a posteriori* approach insofar as it examines historical cases of science to address questions concerning what constitutes good science.

These differences in Carnap's and Kuhn's approaches demonstrate why it is misleading to suggest that their philosophical views are similar on issues of incommensurability. Whereas incommensurability, for Carnap, is a *trivial fact* about certain reconstructed linguistic frameworks (e.g., quantum mechanics cannot be fully reconstructed into the terms of classical mechanics), for Kuhn, incommensurability is a *substantive conclusion* that he reaches through historical analysis. A central aspect of Kuhn's thesis is that proponents of competing paradigms 'work in different worlds' and cannot fully communicate with one another. This aspect of Kuhnian incommensurability is entirely antithetical to the spirit of Carnap's logic of science. The logic of science is motivated precisely to resolve scientific debates by clarifying which disagreements are amenable to meaningful resolution, and which are merely pragmatic. As Carnap consistently reported, he was dismayed by fruitless metaphysical debates and the logic of science is a method for resolving these debates. Hence, incommensurability is a substantive conclusion reached by Kuhn (via historical analysis), while it is a starting point for philosophical resolution (via logical analysis) for Carnap.

There are also reasons for resisting the idea that Carnap and Kuhn share a similar view of theory-choice. Kuhn's (1977) suggestion that paradigm choice inevitably involves the application of shared values that are subjectively interpreted and weighed differently by different scientists is a conclusion that he reaches through an analysis of the history of science. Against philosophers who believe that theory-choice can be 'objective' by restricting theory-choice to the context of justification, Kuhn

---

[7] For a more comprehensive discussion of Kuhn's views on the discovery-justification distinction, see Hoyningen-Huene ([1989] 1993, pp. 245–252; 2006; 13.2.2 of this volume).

suggests that it is illegitimate to separate the contexts of discovery and justification. Kuhn (1977, pp. 326–329) complains that philosophical analyses that confine theory-choice to the context of justification systematically neglect factors that were historically regarded as relevant evidence and they tend to overemphasize arguments that supported the triumphant theory, while neglecting arguments that supported the losing theory. Thus, Kuhn's conclusion that theory choice inevitably involves non-epistemic factors is inferred—in part—on the basis of historical analysis. By contrast, Carnap addresses questions regarding theory choice exclusively in the context of justification. Moreover, while both Kuhn and Carnap can be said to share an instrumentalist (or pragmatic) stance on theory choice,[8] Carnap's instrumentalism is far more robust. Consider differences between their views on the 'choice' between classical mechanics and relativistic physics. Whereas Kuhn ([1962] 1996, Ch. 9) suggests that this was a *forced choice* between incommensurable paradigms that would define the field of physics, Carnap maintains that a decision can appeal to different purposes of physicists (Earman 1993, p. 22). On Carnap's view, each theory is useful for different purposes, e.g., classical mechanics is useful for purposes of measuring and making predictions about objects moving slower than 3×10^8 m/s, while relativistic mechanics is more useful for objects moving faster than 3×10^8 m/s. While Kuhn insists that this revolution was a case of relativistic physics *replacing* classical mechanics, Carnap adopts the more deflationary conventionalist stance that relativistic physics is an instrument that can be freely chosen on pragmatic grounds.

The aforementioned differences vitiate the argument that Carnap and Kuhn share a very similar view of scientific revolutions. While Carnap describes scientific revolutions in terms of a transition from one linguistic framework to another, it is crucial to see that Carnap's idea that a revolution represents a pragmatic choice to adopt a new language is antithetical to Kuhn's view. Kuhn (2000) writes:

> [T]he cognitive importance of language change was for [Carnap] merely pragmatic. One language might permit statements that could not be translated into another, but anything properly classified as scientific knowledge could be both stated and scrutinized in either language, using the same method and gaining the same result.... This aspect of Carnap's position has never been available to me. *Concerned... with the development of knowledge, I have seen each stage in the evolution of a given field as built... upon its predecessors, the earlier stage providing the problems of the stage that followed*. In addition, I have insisted that *some changes in conceptual vocabulary are required for the assimilation and development of the observations, laws, and theories deployed in the later stage*... Given those beliefs, *the process of transition from old state to new becomes an integral part of science, a process that must be understood... to analyze the cognitive basis for scientific beliefs. Language change is cognitively significant for me as it was not for Carnap.* (pp. 227–228, emphasis added)

---

[8] Kuhn ([1962] 1996, Ch. 13, pp. 205–207) adopts an instrumentalist stance on theory-choice insofar as he suggests that, *historically*, paradigms that emerged as victors did so because they had *greater puzzle-solving power*, i.e., they could solve a significant number of puzzles of the previous paradigm and could also solve new puzzles (see Tsou 2006, pp. 216–217).

This passage indicates why Kuhn believes one cannot reduce a scientific revolution to a change in Carnapian linguistic frameworks. For Kuhn, the shift from one paradigm to another is not *merely* a pragmatic choice, but a *naturalistic and historical process* wherein scientists revise, assimilate, and respond to puzzles of the previous paradigm. From Kuhn's perspective, reducing revolutions to a Carnapian external question is to trivialize the problem of scientific change. To properly understand scientific development, one must *historically analyze* how new paradigms emerged out of old ones (Kuhn [1962] 1996, Ch. 1).

5.5 *Structure* and the Demise of Logical Empricism

Reflections on Kuhn's and Carnap's contrasting approaches to philosophy of science help to clarify the historical role played by Kuhn's *Structure* in the decline of logical empiricism (cf. note 6). According to the analysis of this paper, *Structure* primarily contributed to the demise of logical empiricism by offering a concrete and fruitful example of a historically-oriented, bottom-up methodology for philosophy of science that opposed the top-down methodology associated with Carnap and the logical empiricists. This understanding provides reasons for rejecting Irzik and Grünberg's (1995, p. 304) contention that post-positivist philosophy of science associated with Kuhn does not represent a revolutionary departure from Carnapian philosophy of science.

The chief role that *Structure* played in the decline of logical empiricism and emergence of post-positivist philosophy of science was methodological. As discussed in Sect. 5.4, Carnap's logic of science adopts a top-down methodology insofar as it begins with certain *a priori* assumptions about what constitutes good science (e.g., science is non-metaphysical and empirical) and evaluates particular theories on the basis of these criteria. Kuhn's approach, by contrast, adopts a bottom-up methodology insofar as it begins with a historical examination of actual scientific practices and aims to draw philosophical conclusions about what constitutes good science via historical analysis.[9] In this manner, *Structure* offered a novel method for analyzing science that significantly departed from the logical analyses championed by Carnap. By shifting the philosophical unit of analysis away from abstract scientific theories (in the context of justification) and towards historically-situated scientific theories (in the context of discovery), historical analysis—as exemplified in *Structure*—provided an alternative model for doing philosophy of science that would strongly influence subsequent generations of philosophers of science (and science

---

[9] Kuhn's bottom-up methodology can be understood as a *particularist* (as opposed to *generalist*) approach to philosophy of science. From this perspective, *Structure* can be located in a broader tradition of particularist approaches in philosophy (e.g., see Kant [1781] 1998; Wittgenstein [1953] 1958; Sellars [1956] 1997; McDowell 1979; Brandom 1994). For connections between Kuhn's and Wittgenstein's views, see Kindi (1995; 2012) and Sharrock and Reid (2002). As an alternative to this reading of *Structure* as a particularist approach, see Richardson, Chap. 4, this volume.

studies more generally).[10] Kuhn's bottom-up methodology opened avenues for more broadly-focused and interdisciplinary philosophical analyses, and especially analyses that were closely engaged with the history of science. In contrast to the formal analyses of concepts, such as 'confirmation' and 'explanation', offered by logical empiricists, philosophical analyses inspired by *Structure* focused on a broader range of topics, such as the role of experiments in science (Hacking 1983; Franklin 1986; Galison 1987; Chang 2004), the connections between conceptual development in science and research in cognitive science (Giere 1988; Thagard 1988; 1992; Nersessian 2008; Andersen et al. 2006), analogical and model-based reasoning in science (Hesse [1963] 1966; Magnani et al. 1999; Nersessian 2008), and the social dimensions of science (Hull 1988; Longino 1990, 2002; Douglas 2009; Wray 2011).

In addition to presenting an attractive alternative methodological approach for philosophers of science, *Structure* also contributed to the demise of logical empiricism—on a sociological level—by attacking a caricatured 'everyday image of logical positivism' and illegitimately associating it with the cumulative vision of scientific progress rejected in *Structure* (Richardson 2007; Irzik 2012).[11] While the picture of logical empiricism presented by Kuhn in *Structure* was vastly underdeveloped and ultimately misleading, the effect of his work was to stabilize and popularize a simplistic view of logical empiricism as a naïve brand of empiricist foundationalism (Richardson 2007, pp. 359–369), which Kuhn and subsequent philosophers of science could employ as foils for their own arguments.[12] Hence, *Structure* also contributed to the demise of logical empiricism—and the rise of postpositivist philosophy of science—by promulgating and reifying a false image of logical empiricism.

---

[10] In the 1960s and 1970s, Kuhn's *Structure* emerged as the most iconic and influential example of the new historical philosophy of science associated with writers such as Norwood Russell Hanson (1958), Stephen Toulmin (1961), Paul Feyerabend (1975), and Larry Laudan (1977). Retrospectively, these works have jointly been responsible for the 'historical turn' in philosophy of science (Bird 2008). In addition to its influence in philosophy of science, *Structure* had an arguably larger influence in the social sciences, especially among sociologists of science (see Bird 2000, Chap. 7); Kuhn famously repudiated relativist interpretations of his work by proponents of the Strong Programme of the sociology of scientific knowledge (Kuhn 2000, Chap. 5). For further discussion of Kuhn's relation to the sociology of science and the Strong Programme, see K. Brad Wray (Chap. 12) in this volume.

[11] With characteristic honesty, Kuhn admitted that he had not read any of the mature works of Carnap when he was writing *Structure* (see Borradori [1991] 1994, p. 153; Kuhn 2000, pp. 227, 305–306; Irzik 2012, appendix). Wray (2013) points out that Kuhn likely did not feel the need to read later positivist works since he was well-acquainted with Quine's ([1951] 1980) critiques of Carnap, while Kuhn and Quine were colleagues at the Harvard Society of Fellows (see Kuhn 2000, p. 279).

[12] Besides *Structure*, Quine's influential criticisms of Carnap (Quine [1951] 1980, 1969) undoubtedly served to stabilize the image of logical empiricism as an impoverished project in empiricist foundationalism (see Reisch 2005, pp. 3–5). For criticisms of Quine's presentation of Carnap in the context of the Carnap-Quine analyticity debates, see Creath (1991), Stein (1992), and Friedman (1999, 2001).

At this point, what is correct and incorrect in the revisionist view can be stated with clarity. One of the motivations of the revisionist view is to urge that Carnap is a much more methodologically sophisticated philosopher of science than is typically thought and that many of the alleged weaknesses of his view are simply misplaced.[13] I am largely sympathetic with this aspect of the revisionist argument. A large part of the narrative surrounding the received view on Carnap and Kuhn is that one of Kuhn's chief achievements in *Structure* was to demonstrate the methodological flaws of logical empiricism by highlighting the importance of issues such as the theory-ladenness of observation, the underdetermination of theories by evidence, and the non-epistemic aspects of science. The received view on Carnap and Kuhn is incorrect in assuming that Carnap was unaware of, or insensitive to, these issues; in fact, many of Carnap's views were motivated by precisely these issues. Hence, the revisionist analysis is correct to point out that Carnap and Kuhn shared many of the same methodological assumptions and that there are similarities in the way that they understood the epistemological structure of scientific theories. However, the revisionist view is incorrect in drawing the stronger conclusion that Carnap's and Kuhn's shared assumptions render their philosophies closely aligned. I have argued that the similarities that revisionist analyses highlight are superficial. It is only by ignoring the fundamental differences in Carnap's and Kuhn's broader philosophical projects that this conclusion can be drawn.

The analysis of this chapter also provides a corrective to the received view on the Carnap-Kuhn relationship. What is correct in the received view is that Kuhn's *Structure* ushered in a new style of doing philosophy of science, which significantly differed from Carnap's favored logical approach for analyzing science. Compared to Carnap, Kuhn's approach was characterized by a bottom-up approach to analyzing scientific theories and an emphasis on integrating the history of science into philosophical analyses. However, the received view is incorrect in maintaining Carnap's and Kuhn's views are diametrically opposed and that the primary achievement of *Structure* was to demonstrate the false assumptions of positivist philosophy of science. Rather, Kuhn's historical significance in the history of twentieth century philosophy of science was to change the focus of philosophy of science (e.g., from formal analyses of confirmation to issues concerning the nature of scientific change) and to change the favored methodological tools that philosophers of science employed to analyze science (i.e., from formal tools to historical resources). As I have argued, Kuhn's chief methodological achievement was to offer an alternative bottom-up

---

[13] This aspect of the revisionist argument is part of the larger movement of historical scholarship on logical empiricism (e.g., see Coffa 1991; Cartwright et al. 1996; Giere and Richardson 1996; Nemeth and Stadler 1996; Richardson 1998; Friedman 1999; Hardcastle and Richardson 2003; Stadler 2003; Awodey and Klein 2004; Okruhlik 2004; Reisch 2005; Carus 2007; Friedman and Creath 2007; Richardson and Uebel 2007; Uebel 2007; Creath 2012), which has revealed both the great complexity of thought as well as heterogeneity within logical empiricism.

and historical approach to analyzing scientific theories, which was opposed to the top-down and logical approach to analyzing theories championed by Carnap. In this sense, *Structure* undoubtedly represents a revolutionary departure from positivist philosophy of science, as exemplified in Carnap's work.

5.6 Conclusion

In this chapter, I offered reasons for rejecting the conclusion that Carnap's and Kuhn's philosophies are closely aligned. In particular, I argued against the revisionist conclusions that Carnap and Kuhn share similar views on incommensurability, theory-choice, and scientific revolutions. Moreover, I argued that fundamental differences between their styles of philosophy of science pertained to their preferred methods of analyzing science (i.e., logical versus historical analysis), their stance on the context of discovery/ justification distinction, and the relative importance they place on accurately reconstructing science. According to this analysis, the primary role that *Structure* played in the demise of logical empiricism was to offer a novel bottom-up methodological approach for analyzing scientific theories that shifted subsequent generations of philosophers of science away from the top-down approach espoused by Carnap.

In the context of contemporary philosophy of science, Carnap and Kuhn can be regarded as model representatives of two distinctive traditions of doing philosophy of science. In the Carnapian tradition are philosophers whose analyses investigate science exclusively in the context of justification and favor formal methods. In the Kuhnian tradition are philosophers whose analyses are closely engaged with the history of science and aim to draw philosophical conclusions from historical case studies. While these two styles of philosophy of science need not be regarded as inherently incompatible, it is mistaken to think that these traditions spurned by Carnap and Kuhn do not constitute fundamentally different ways of doing philosophy of science.

Acknowledgments I am grateful to Alan Richardson, Vasso Kindi, Christian Damböck, Trevor Pearce, Scott Edgar, Ian Hacking, Bill Wimsatt, Paul Hoyningen-Huene, Chris Pincock, Morgan Harrop, Philip Hanson, Gürol Irzik, George Reisch, Greg Frost-Arnold, David Marshall Miller, David Alexander, Uljana Feest, Thomas Uebel, and Matteo Collodel for helpful comments and suggestions. Earlier drafts of this paper were presented at the ninth biennial meeting of the International Society for the History of Philosophy of Science (HOPOS) at University of King's College, Halifax, NS, June 2012; the twenty-third biennial meeting of the Philosophy of Science Association (PSA) in San Diego, CA, November 2012; and the third annual Philosophy Alumni Conference at Simon Fraser University, Burnaby, BC, March 2013. I am grateful for feedback that I received on these occasions. Special thanks to Harrop (my commentator at the SFU alumni conference) for his detailed and extensive written comments.

References

Andersen, H., P. Barker, and X. Chen. 2006. *The cognitive structure of scientific revolutions.* Cambridge: Cambridge University Press.

Awodey, S., and Klein, C., eds. 2004. *Carnap brought home: The view from Jena.* Chicago: Open Court.

Axtell, G. S. 1993. In the tracks of the historicist movement: Re-assessing the Carnap-Kuhn Connection. *Studies in History and Philosophy of Science* 24 (1): 119–146.

Bird, A. 2000. *Thomas Kuhn.* Princeton: Princeton University Press.

Bird, A. 2002. Kuhn's wrong turning. *Studies in History and Philosophy of Science* 33 (3): 443–463.

Bird, A. 2004. Kuhn, naturalism, and the positivist legacy. *Studies in History and Philosophy of Science* 35 (2): 337–356.

Bird, A. 2008. The historical turn in the philosophy of science. In *Routledge companion to the philosophy of science,* ed. S. Psillos and M. Curd, 67–77. Abingdon: Routledge.

Borradori, G. 1991/1994. *The American philosopher: Conversations with Quine, Davidson, Putnam, Nozick, Danto, Rorty, Cavell, MacIntyre, and Kuhn.* Trans.: R. Crocitto. Chicago: University of Chicago Press. Originally published as *Conversazioni americane con W. O. Quine, D. Davidson, H. Putnam, R. Nozick, A. C. Danto, R. Rorty, S. Cavell, A. MacIntyre, Th. S. Kuhn.* Gius: Laterza & Figli.

Brandom, R. B. 1994. *Making it explicit: Reasoning, representing, and discursive commitment.* Cambridge: Harvard University Press.

Carnap, R. 1931/1934. *The unity of science.* Trans.:M. Black. London: Kegan Paul, Trench Trubner and Co. Originally published as: Die physikalische Sprache als Universalsprache der Wissenschaft. *Erkenntnis* 2 (5/6): 432–465.

Carnap, R. 1931/1959. The elimination of metaphysics through logical analysis of language. In *Logical positivism,* ed. A. J. Ayer, 60–81. New York: The Free Press. Originally published as: Überwindung der Metaphysik durch logische Analyse der Sprache. *Erkenntnis* 2(4): 219–241.

Carnap, R. 1934/1937. *The logical syntax of language.* Trans.: A. Smeaton. London: Kegan Paul, Trench Trubner and Co. Originally published as: *Logische Syntax der Sprache.* Vienna: Verlag von Julius Springer.

Carnap, R. 1936/1949. Truth and confirmation. In *Readings in philosophical analysis,* ed. H. Feigl and W. Sellars, 119–127. New York: Appleton-Century-Crofts. Originally published as: Wahrheit und Bewährung. *Actes du Congrès international de philosophie scientifique,* vol. 4, 18–23. Paris: Hermann & Cie, éditeurs.

Carnap, R. 1936. Testability and meaning. *Philosophy of Science* 3 (4): 419–471.

Carnap, R. 1937. Testability and meaning-continued. *Philosophy of Science* 4 (1): 1–40.

Carnap, R. 1950/1956. Empiricism, semantics, and ontology. In *Meaning and necessity: A study in semantics and modal logic,* ed. R. Carnap, 2nd ed. 205–221. Chicago: University of Chicago Press.

Carnap, R. 1956. The methodological character of theoretical concepts. In *The foundations of science and the concepts of psychology and psychoanalysis,* ed. H. Feigl and M. Scriven, 38–76. Minneapolis: University of Minnesota Press. (Minnesota Studies in the Philosophy of Science, vol. 1).

Carnap, R. 1963. Replies and systematic expositions. In *The philosophy of Rudolf Carnap: Library of living philosophers,* ed. P. A. Schilpp, vol. 11, 859–1013. LaSalle: Open Court.

Cartwright, N., J. Cat, L. Fleck and T. E. Uebel. 1996. *Otto Neurath: Philosophy between science and politics.* Cambridge: Cambridge University Press.

Carus, A. W. 2007. *Carnap and twentieth-century thought: Explication as enlightenment.* Cambridge: Cambridge University Press.

Chang, H. 2004. *Inventing temperature: Measurement and scientific progress.* Oxford: Oxford University Press.

Coffa, J. A. 1991. *The semantic tradition from Kant to Carnap: To the Vienna station,* ed. L. Wessels. Cambridge: Cambridge University Press.

Creath, R. 1991. Every dogma has its day. *Erkenntnis* 35 (1–3): 347–389.

Creath, R. ed. 2012. *Rudolf Carnap and the legacy of logical empiricism*. Dordrecht: Springer. (Vienna Circle Institute Yearbook, vol. 16).

Demopoulos, W. 2003. On the rational reconstruction of our theoretical knowledge. *British Journal for the Philosophy of Science* 54 (3): 371–403.

Demopoulos, W. 2007. Carnap on the rational reconstruction of scientific theories. In *The Cambridge companion to Carnap*, ed. M. Friedman and R. Creath, 248–272. Cambridge: Cambridge University Press.

DiSalle, R. 2002. Reconsidering Kant, Friedman, logical positivism, and the exact sciences. *Philosophy of Science* 69 (2): 191–211.

Douglas, H. E. 2009. *Science, policy, and the value-free ideal*. Pittsburgh: University of Pittsburgh Press.

Earman, J. 1993. Carnap, Kuhn, and the philosophy of scientific methodology. In *World changes: Thomas Kuhn and the nature of science*, ed. P. Horwich, 9–36. Cambridge: MIT Press.

Feyerabend, P. K. 1970. Consolations for the specialist. In *Criticism and the growth of knowledge*, ed. I. Lakatos and A. Musgrave, 197–230. Cambridge: Cambridge University Press.

Feyerabend, P. K. 1975. *Against method: Outline of an anarchistic theory of knowledge*. London: New-Left Books.

Franklin, A. 1986. *The neglect of experiment*. Cambridge: Cambridge University Press.

Friedman, M. 1998. On the sociology of knowledge and its philosophical agenda. *Studies in History and Philosophy of Science* 29 (2): 239–271.

Friedman, M. 1999. *Reconsidering logical positivism*. Cambridge: Cambridge University Press.

Friedman, M. 2001. *Dynamics of reason*. Stanford: CSLI Publications.

Friedman, M. 2003. Kuhn and logical empiricism. In *Thomas Kuhn*, ed. T. Nickles, 19–44. Cambridge: Cambridge University Press.

Friedman, M., and R. Creath, eds. 2007. *The Cambridge companion to Carnap*. Cambridge: Cambridge University Press.

Galison, P. 1987. *How experiments end*. Chicago: University of Chicago Press.

Galison, P. L. 1995. Context and constraints. In *Scientific practice: Theories and stories of doing physics*, ed. J. Z. Buchwald, 13–41. Chicago: University of Chicago Press.

Gattei, S. 2008. *Thomas Kuhn's "linguistic turn" and the legacy of logical empiricism: Incommensurability, rationality and the search for truth*. Aldershot: Ashgate.

Giere, R. N. 1985. Philosophy of science naturalized. *Philosophy of Science* 52 (3): 331–356.

Giere, R. N. 1988. *Explaining science: A cognitive approach*. Chicago: University of Chicago Press.

Giere, R. N., and A. W. Richardson, eds. 1996. *Origins of logical empiricism*. Minneapolis: University of Minnesota Press. (Minnesota Studies in the Philosophy of Science, vol. 16).

Hacking, I. 1983. *Representing and intervening: Introductory topics in the philosophy of science*. Cambridge: Cambridge University Press.

Hanson, N. R. 1958. *Patterns and discovery: An inquiry into the conceptual foundations of science*. Cambridge: Cambridge University Press.

Hardcastle, G. L., and A. W. Richardson, eds. 2003. *Logical empiricism in North America*. Minneapolis: University of Minnesota Press. (Minnesota Studies in the Philosophy of Science, vol. 18).

Hempel, C. G. 1965. Empiricist criteria of cognitive significance: Problems and changes. In *Aspects of scientific explanation and other essays in the philosophy of science*, ed. C. G. Hempel, 101–119. New York: The Free Press.

Hesse, M. 1963/1966 *Models and analogies in science*, 2nd ed. Notre Dame: University of Notre Dame Press.

Hoyningen-Huene, P. 1989/1993. *Reconstructing scientific revolutions: Thomas S. Kuhn's philosophy of science*, Trans.: A. T. Levine. Chicago: University of Chicago Press. Originally published as: *Die Wissenschaftsphilosophie Thomas S. Kuhns: Rekonstruktion und Grundlagenprobleme*. Braunschweig: Vieweg.

Hoyningen-Huene, P. 2006. Context of discovery versus context of justification and Thomas Kuhn. In *Revisiting discovery and justification: Historical and philosophical perspectives on the context distinction*, ed. J. Shickore and F. Steinle, 119–131. Dordrecht: Springer.

Hull, D. 1988. *Science as a process: An evolutionary account of the social and conceptual development of science*. Chicago: University of Chicago Press.

Irzik, G. 2002. Carnap and Kuhn: A belated encounter. In *In the scope of logic, methodology and philosophy of science*, vol. 2, ed. P. Gärdenfors, J. Woleński, and K. Kijania-Placek, 603–620. Dordrecht: Kluwer.

Irzik, G. 2003. Changing conceptions of rationality: From logical empiricism to postpositivism. In *Logical empiricism: Historical & contemporary perspectives*, ed. P. Parrini, W. Salmon, and M. H. Salmon, 325–346. Pittsburgh: University of Pittsburgh Press.

Irzik, G. 2012. Kuhn and logical empiricism: Gaps, silences, and tactics in SSR. In *Kuhn's The structure of scientific revolutions revisited*, ed. V. Kindi and T. Arabatzis, 15–40. New York: Routledge.

Irzik, G., and T. Grünberg. 1995. Carnap and Kuhn: Arch enemies or close allies? *British Journal for the Philosophy of Science* 46 (3): 286–307.

Irzik, G., and T. Grünberg. 1998. Whorfian variations on Kantian themes: Kuhn's linguistic turn. *Studies in History and Philosophy of Science* 29 (2): 207–221.

Jeffrey, R. 1994. Carnap's voluntarism. In *Logic, methodology, and philosophy of science IX: Proceedings of the ninth International Congress of Logic, Methodology, and Philosophy of Science, Uppsala, Sweden, August 7–14, 1991*, ed. D. Prawitz, B. Skyrms, and D. Westerstål, 847–866. Amsterdam: Elsevier.

Kant, I. 1781/1998. *Critique of pure reason*, Trans.: P. Guyer and A. W. Wood. Cambridge: Cambridge University Press. Originally published as: *Critik der reinen Vernunft*. Riga: Verlegts Johann Friedrich Hartknoch.

Kindi, V. P. 1995. Kuhn's The structure of scientific revolutions revisited. *Journal for General Philosophy of Science* 26 (1): 75–92.

Kindi, V. 2005. The relation of history of science to philosophy of science The structure of scientific revolutions and Kuhn's later philosophical work. *Perspectives on Science* 13 (4): 495–530.

Kindi, V. 2012. Kuhn's paradigms. In *Kuhn's The structure of scientific revolutions revisited*, ed. V. Kindi and T. Arabatzis, 91–111. New York: Routledge.

Kuhn, T. S. 1962/1996. *The structure of scientific revolutions*, 3rd ed. Chicago: University of Chicago Press.

Kuhn, T. S. 1977. Objectivity, value judgment, and theory choice. In *The essential tension: Selected studies in scientific tradition and change*, ed. T. S. Kuhn, 320–339. Chicago: University of Chicago Press.

Kuhn, T. S. 2000. *The road since Structure: Philosophical essays, 1970–1993, with an autobiographical interview*, ed. J. Conant and J. Haugeland Chicago: University of Chicago Press.

Lange, M. 2004. Review essay on Dynamics of reason by Michael Friedman. *Philosophy and Phenomenology Research* 68 (3): 702–712.

Laudan, L. 1977. *Progress and its problems: Towards a theory of scientific growth*. Berkeley: University of California Press.

Longino, H. E. 1990. *Science as social knowledge: Values and objectivity in scientific inquiry*. Princeton: Princeton University Press.

Longino, H. E. 2002. *The fate of knowledge*. Princeton: Princeton University Press.

Magnani, L., N. J. Nersessian, and P. Thagard, eds. 1999. *Model-based reasoning in scientific discovery*. New York: Kluwer Academic/Plenum Publishers.

McDowell, J. 1979. Virtue and reason. *The Monist* 62 (3): 331–350.

McGuire, J. E. 1992. Scientific change: Perspectives and proposals. In *Introduction to the philosophy of science: A text by members of the department of history and philosophy of science of the University of Pittsburgh*, ed. M. H. Salmon et al. 132–178. Indianapolis: Hackett.

Mladenović, B. 2007. Muckraking in history: The role of the history of science in Kuhn's philosophy. *Perspectives on Science* 15 (3): 261–294.

Nemeth, E., and F. Stadler, eds. 1996. *Encyclopedia and utopia: The life and work of OttoNeurath (1882–1945)*. Dordrecht: Kluwer (Vienna Circle Institute Yearbook, vol. 4).

Nersessian, N. J. 2008. *Creating scientific concepts*. Cambridge: MIT Press.

Okruhlik, K. 2004. Logical empiricism, feminism, and Neurath's auxiliary motive. *Hypatia* 19 (1): 48–72.

Pincock, C. 2012. *Mathematics and scientific representation*. Oxford: Oxford University Press.

Pinto de Oliveira, J. C. 2007. Carnap, Kuhn, and revisionism: On the publication of Structure in the Encyclopedia. *Journal for General Philosophy of Science* 38 (1): 147–157.

Preston, J. 2004. Bird, Kuhn, and positivism. *Studies in History and Philosophy of Science* 35 (2): 327–335.

Quine, W. V. 1951/1980. Two dogmas of empiricism. In *From a logical point of view: Nine logico-philosophical essays*, ed. W. V. Quine, 2nd ed. 20–46. Cambridge: Harvard University Press. Originally published in *The Philosophical Review* 60 (1): 20–43.

Quine, W. V. 1969. Epistemology naturalized. In *Ontological relativity and other essays*, ed. W. V. Quine, 69–90. New York: Columbia University Press.

Reisch, G. A. 1991. Did Kuhn kill logical positivism? *Philosophy of Science* 58 (2): 264–277.

Reisch, G. A. 2005. *How the cold war transformed philosophy of science: To the icy slopes of logic*. Cambridge: Cambridge University Press.

Richardson, A. W. 1998. *Carnap's construction of the world: The Aufbau and the emergence of logical empiricism*. Cambridge: Cambridge University Press.

Richardson, A. W. 2002. Narrating the history of reason itself: Friedman, Kuhn, and a constitutive a priori for the twenty-first century. *Perspectives on Science* 10 (3): 253–274.

Richardson, A. 2007. That sort of everyday image of logical empiricism: Thomas Kuhn and the decline of logical empiricist philosophy of science. In *The Cambridge companion to logical empiricism*, ed. A. Richardson and T. Uebel, 346–370. Cambridge: Cambridge University Press.

Richardson, A., and T. Uebel, eds. 2007. *The Cambridge companion to logical empiricism*. Cambridge:Cambridge University Press.

Sellars, W. 1956/1997. *Empiricism and the philosophy of mind*. Cambridge: Harvard University Press. Originally published in *Foundations of Science and the concepts of psychology and psychoanalysis*, ed. H. Feigl and M. Scriven, 253–329. Minneapolis: University of Minnesota Press (Minnesota Studies in the Philosophy of Science, vol. 1).

Sharrock, W., and R. Reid. 2002. *Kuhn: Philosopher of scientific revolution*. Cambridge: Polity Press.

Stadler, F., ed. 2003. *The Vienna Circle and logical empiricism: Re-evaluation and future perspectives*. New York: Kluwer. (Vienna Circle Institute Yearbook, vol. 10).

Stein, H. 1992. Was Carnap entirely wrong, after all? *Synthese* 93 (1–2): 275–295.

Suppe, F. 1974/1977. The search for philosophic understanding of scientific theories. In *The structure of scientific theories,* ed. F. Suppe, 2nd ed., 3–241. Urbana: Illinois University Press.

Thagard, P. 1988. *Computational philosophy of science*. Cambridge: MIT Press.

Thagard, P. 1992. *Conceptual revolutions*. Princeton: Princeton University Press.

Toulmin, S. 1961. *Foresight and understanding: An enquiry into the aims of science*. Bloomington: Indiana University Press.

Tsou, J. Y. 2003. Critical notice: A role for reason in science. *Dialogue: Canadian Philosophical Review* 42 (3): 573–398.

Tsou, J. Y. 2006. Genetic epistemology and Piaget's philosophy of science: Piaget vs. Kuhn on scientific progress. *Theory & Psychology* 16 (2): 203–224.

Uebel, T. 2007. *Empiricism at the crossroads: The Vienna circle's protocol sentence debate*. Chicago: Open Court.

Uebel, T. 2011. Carnap and Kuhn: On the relation between the logic of science and the history of science. *Journal for General Philosophy of Science* 42 (1): 129–140.

Wittgenstein, L. 1953/1958. Philosophical investigations, 2nd ed., G. E. Anscombe, trans. Oxford: Basil Blackwell.
Wray, K. B. 2011. *Kuhn's evolutionary social epistemology*. Cambridge: Cambridge University Press.
Wray, K. B. 2013. Review of Vasso Kindi and Theodore Arabatzis eds., Kuhn's The structure of scientific revolutions revisited. *Notre Dame Philosophical Reviews*. http://ndpr.nd.edu/news/38265-kuhn-s-the-structure-of-scientific-revolutions-revisited/. Accessed 13 March 2013.

Chapter 6
The Rationality of Science in Relation to its History

Sherrilyn Roush

6.1 Introduction

Kuhn's *Structure of Scientific Revolutions* richly displayed the relevance of historical considerations to questions of what it is rational for scientists to believe and why. He did not, as some have maintained, think his interpretation threatened to make scientists' beliefs look irrational, but his work did make an approach to the issue of rationality via rules of Scientific Method seem cartoonish. For good or ill, though today there are more Bayesians than falsificationists and we tend to avoid that two-word proper name "Scientific Method" that now seems so naïve, philosophers have not stopped investigating methodological rules and principles of rationality. And many philosophers continue to take the relevance of the history of science to be as a pool of cases to be used to illustrate and test our views of the general rules of rationality, which involves nothing essentially historical.

I will not apologize for either proposals of general theories of the rationality of science, or the effort to keep them on topic by means of examples (Donovan et al. 1992). While historians since Kuhn tend to be suspicious of generalizations about all science, ironically Kuhn himself proposed an inner logic as general as any methodologist's to explain the rationality of scientists' beliefs and behavior, in terms of the cycle: paradigm—normal science—anomaly—crisis— revolution. But Kuhn's generalizations arose *out of* attention to the specifics, the differences over time, the challenges, conceptual and practical, of getting a theory to say anything about the

S. Roush (✉)
Department of Philosophy, King's College London, Strand, WC2R 2LS London, UK
e-mail: sherrilyn.roush@kcl.ac.uk; sherri.roush@gmail.com

© Springer International Publishing Switzerland 2015
W. J. Devlin, A. Bokulich (eds.), *Kuhn's Structure of Scientific Revolutions—50 Years On,*
Boston Studies in the Philosophy and History of Science 311,
DOI 10.1007/978-3-319-13383-6_6

world, out of what David Hollinger has called the "quotidian".[1] Though Kuhn was a philosopher in my view, he had a historian's eye, which most philosophers lack. In contrast with Kuhn's direction, with philosophers today, including myself, the path of discovery of generalizations about the rationality of science often goes from general epistemological considerations—such as the worthiness of betting in line with the axioms of probability— down, by deduction, to slightly more specific principles intended to explain scientists' behavior, such as that surprising evidence has more confirming power. Such principles are tested against cases antecedently regarded as sound or unsound, but the fact that some of those cases are from the past is not, per se, significant.

However there is another prominent pattern among philosophers of taking the history of science as relevant to its rationality that appears to flow in Kuhn's direction from historical facts to generalizations. For example, in some arguments for scientific realism—the view that we have reason to believe successful theories (or their essential structures) are 'true-ish', rather than just good predictors of observables—putative facts about the history of science are taken as premises. The predictive success, retention, and convergence of opinion over historical time of some theories has been taken to be a reason to think those theories are true-ish. Another case of this pattern, the one that will concern me here, is found in the pessimistic induction (PI), which has nagged at the consciences of philosophers, laypersons, and even some scientists since the 1980s. The premise of this argument is also the putative track record of science, but here it is the failures: a high proportion of past scientists' theories were successful, but by our lights false in what they said about unobservable entities, as in the cases of phlogiston and the luminiferous ether, goes the argument. Induction, broadly construed, is the method of science, and that invites an induction over this sad record to the conclusion that it is not rational to have much confidence that our current successful theories accurately represent the world we cannot see. The history of science thus presents a challenge to the rationality of endorsing our scientists' hypotheses about unobservables at what one might call face value.

The historical claim merely that there have existed successful theories whose claims about unobservables were false is not generally disputed because it is weak, too weak to establish a worrying pessimistic conclusion. That there were a few such cases or a small proportion of the total cases would give us at most a conclusion that our theories have some chance of being false, and we already knew that. To make the historical claim strong enough for trouble, it would have to be that success with observables is not even a mark of truth. This would require knowing the proportion of false theories in the pools of the successful and unsuccessful theories, but it has been argued that we cannot know these two base rates without begging the question. (Lewis 2001; Magnus and Callender 2004).

---

[1] David Hollinger has recently discussed this theme in a lecture honoring Kuhn at MIT in December 2012. He originally developed it in a paper (Hollinger 1973) that was much appreciated by Kuhn, who distributed reprints of it to thirty of his acquaintants.

However, as I will argue, even if the pessimist could defend appropriate claims about the base rates, his argument would not enjoy smooth sailing. Those rates go to secure the premises of the pessimist's argument, which are general facts about history, such as that many successful theories have been false, or that our predecessors have often failed to conceive of conceivable hypotheses. The last step of the pessimist's argument, the "action" one might say, begins with the generalizations, not the particular cases that they are generalizations about. Thus though the pessimist does proceed *from* claims about history, the specifics of the failures he points to—the sort of thing historians have eyes for—quickly become apparently irrelevant to the argument.

This focus is natural—the conclusion that we should be less confident in our theories is general, so it is presumed that the relevant premises will be general too—but I will argue here that this presumption is false. In an induction, including a PI, for the most normative, general, epistemological reasons there is no point at which the specifics of cases cease to be relevant to the argument. I will use this to argue that even if we could establish damning generalizations about our predecessors, it would not be sufficient for the pessimist's argument to touch the confidences in our own theories that we have arrived at by the usual means of the particular science. On the basis of general points about induction I will argue that novelty and discontinuity over time, particularly a novelty of method that tends to go unappreciated, saves our science from the PI. Ironically, the relevance of novelty is that it renders most failures of the past irrelevant to the rationality of embracing our own theories. The novelty that has the potential for this effect on the pessimist's argument is revealed at every more specific level of description of the past and present cases.

I will conclude that no form of the PI can succeed in giving us doubts over and above those that competent scientists already address in their day-to-day work on particular hypotheses. I will give a diagnosis of the hold the pessimistic argument continues to have over us, and explain this problem as the dual of a familiar rationality problem that Kuhn gave to us by introducing the notion of paradigm shifts.

6.2   The Objective

To defend our science against the PI is not to make any positive, general argument about what kind of truths science does get us, but only to show that the pessimist's negative conclusion is not supported by his evidence. Also, for the pessimist to succeed requires him to give us a reason to withdraw confidence in our particular hypotheses, such as those about the mechanism of chemical mutation in E. coli. DNA, and the convection currents and composition of the Sun.[2] This is what is at stake in the argument. Scientists have evidence and arguments for such hypotheses, apportioning

---

[2] When I write "particular" hypotheses I do not mean singular propositions—the mechanism of chemical mutation in E. coli. is a phenomenon with more than one instance—but rather propositions investigated in particular sciences.

their confidences in these hypotheses to what they judge to be the strength of their evidence for them, and the pessimist's argument must show why those things the good scientist already does are not sufficient to justify their confidences in particular hypotheses if he is going to show anything troubling.

One way to do that would be to offer counterevidence about the composition or convection currents of the Sun, or to offer meta-arguments casting doubt on the design of the *E. coli.* experiments. We know that providing evidence against particular scientific hypotheses or experimental designs is not what the pessimist is doing; he is providing an argument based on generalizations. That the PI is reflective does not excuse scientists from addressing it, of course. Scientists reflect on their procedures and arguments every day, but how reflective they are obligated to be is limited by the quality of the meta-level objections. So, the pessimist has the burden to give an argument of good quality that is distinct from those scientists already address, if we are to have a special problem that derives from the historical record. This, I will claim, is what he cannot do.

6.3 Induction

I will take the pessimist at his word that what he is doing is an induction.[3] The term "induction" includes any ampliative inference, one in which the conclusion contains more information than the premises, in which it is logically possible for the premises to be true and the conclusion false, whatever more particular form that inference takes. The argument I will make is not restricted to next-case induction or generalization, but applies equally well to, for example, inference to causes. My argument, which rests on a claim that relevance demands similarity, does not hold generally for inference to the best explanation, or abduction, but this is not a form of argument the pessimist is using, or I think can use.[4] Given all of this, I will use the term "induction" for the forms of ampliative inference for which my assumptions hold, in the hopes that any kind of ampliative inference for which my claim about similarity and relevance does not hold is also, like abduction, one there is no way for the pessimistic inductivist to exploit.

The first point is that induction needs a similarity base, a similarity between the subjects of the premises and conclusion that makes the premises relevant to the

---

[3] Larry Laudan is usually credited with the first PI (Laudan 1981) but his confutation of realism was not an induction. His argument took the form of historical counterexamples to strong realist claims of a sort I am not defending, for example that empirical success is a mark of truth. The first adumbration of the PI argument seems to have been given by Poincare (1905), who did not however develop or endorse it.

[4] It can be used to defend anti-realism more broadly, of course, in an argument proposing to explain the success of science without appealing to the truth of theories. I only say it does not look helpful in a pessimistic induction over history.

conclusion.[5] Prima facie, we may infer that all swans are white from the fact that all we have seen are white because our evidence and conclusion both ascribe the projected property to swans. By contrast we would not even consider inferring from the fact that all swans we have seen are white that all paper towels are white. Even if all paper towels are white it is not a claim supported by evidence concerning swans.

Secondly, even if there is a similarity base, as with the old swan example, an induction is not justified if there is a known property P that is plausibly relevant to the conclusion property, and P is not uniform between data- and target- populations. In the swan case there is such a property because the habitats of the swans you have seen may easily be different from the habitats of some swans you are projecting to, and there is often color variation within species in different habitats. Such a property P provides what Hans Reichenbach called a "cross-induction".[6] For an induction to be justified, the similarity base must not be undermined by available evidence of a further property of the subjects (here distinct habitats) that is plausibly relevant to the presence or absence of the projected property (here white color).[7]

Often, cross-inductions operate under the surface. A smart-aleck could point out that swans and paper towels do have a similarity base; they both occur in the United States, for example. What is wrong with an induction based on that similarity is that there are further properties of swans and paper towels that are plausibly relevant to color, such as that one is an animal and the other a cleaning item. Another way to think of the situations where cross-inductions are appropriate, and one that will be helpful here, is that under-description of the evidence and the target has concealed the irrelevance of the evidence to the conclusion. Sometimes, we know and apply a fuller description automatically and unconsciously, as with the paper towels; sometimes we discover the further properties for cross-induction in the course of time, as with Europeans discovering black swans in Australia; and sometimes we already know of the further properties but have failed to take them into account. The latter is the case with the PI.

---

[5] This point goes through for inference to causes. When we infer that X causes Y, we do it on the basis of information about things that have property X and have or lack property Y. The evidence and conclusion are about things similar with respect to X.

[6] Cross-induction is an alternative way of describing the non-monotonicity, or erodability, of ampliative inference, namely that, in contrast to deduction, addition of evidence to the premises can undermine the legitimacy of the inference. Epistemologists call such an additional piece of evidence a "defeater" or "underminer".

[7] One might want to strengthen the requirement for justification by weakening the qualifier "available". It seems that there are some examples in which justification can be undermined by further evidence even if one does not possess that evidence (Harman 1980). That stronger requirement is not necessary for my argument, and indeed would weaken the force of my conclusion, since it would weaken the requirement for a cross-induction.

6.4   Similarity Between Past and Present Science

The pessimist needs a similarity base between past and present science to make our predecessors' failures relevant to what we have a right to believe about the world. The pessimist would give us a challenge if he could convince us that we have less right to confidence than we think we do at the object level, at the level of beliefs about particular unobservable matters such as the mechanism of replication of the MERS-CoV virus, and this is the kind of implication the pessimist advertises. Can he find his similarity base at this object level?

Similarity of this sort would be similarity in the content of theories or evidence. The PI is often intended to go way back and across subject matters, even to the theory of crystalline spheres and the theory of bodily humors. But the theory of bodily humors is not similar to the theory of quantum mechanics, not even in its subject matter, much less in its particular claims about its subject. And vast differences in the content of theories—what they claim about the world—is relevant to whether the theories are true.

Sometimes, though, the subject matter of past science is the same as that of ours. Newton's mechanics was declared universal in its scope, so it is not just the part of Newton's theory that we still think is approximately true that is the same subject matter as relativity and quantum mechanics. We do have a different particular theory from that strictly false Newtonian one so we might think that spoils the pessimism. But even in revolutions, physicists do not mutilate where it is not necessary, so there are similarities too, for example, the ones that structural realists argue are retained over the history of modern physics. There are also similarities in the evidence we have and they had for the similar contents because we retain that, too. So what if the similarity base is this: that part of the *content* of scientists' claims about the world that is similar?

This will not work, because if we think through similarity of content in theories and evidence carefully, we will see that the pessimist has a dilemma. Consider, first, the cases where our predecessors' theories were similar to ours in content. Those theories were either true or false. If their theories were false and we retained the false parts in our theories, then our theories are false too, but that is not an induction over history. If their theories were true and ours are similar in the respects that are true, then that is not grounds for pessimism. Similarity of the kind considered does bring relevance, but it does not support a PI.

Second, consider the cases where our predecessors' evidence was similar to ours in content. Any evidence is either supportive, counter-evidence, or irrelevant to a given theory of ours. If it is irrelevant, then it does not matter to how confident we should be in our theories. If our predecessors' evidence is supportive of our theories then that is good for our confidence. If it is counter-evidence, then it is reason to think our theories are false. But then our theories are seen as false because of particular counter-evidence to them, perhaps discovered by doing history, and not because of an induction over the history of science.

Thus, this object-level content strategy does not succeed. The argument reduces to something that either is not an induction or is not pessimistic. If these points are obvious, that is good for my argument. I have excavated them in order to show that the pessimist has no options at the object level. It is not just that the PI is often called and pursued as a meta-induction but that because there is no appropriate similarity at the object level, the argument cannot both be successful and avoid that ascent to the meta-level.

The similarity at the basis of the apparently powerful argument must be a more general one between our predecessors and ourselves as investigators. We are doing the same thing that they did in some important sense. So how can we expect a different result? In particular, we are all doing science, and justified relative to the evidence that we have. Suddenly the theories of crystalline spheres and bodily humors seem relevant again. Our predecessors were unreliable in getting true theories. We are like them, so we are likely also often wrong. That justifiedness that they had and we have must of course be similar for the induction to proceed and that, in my view, is the weak link in the argument, which I will come back to.

But first, note that inducing over this property of justifiedness and to the property of unreliability—being often wrong—makes it a meta-induction in a precise sense. We are at the second order, meaning that the properties in question are properties of the scientist's beliefs, not of the world which it is her primary aim to form beliefs about. In performing the pessimist's argument on ourselves, we are managing our beliefs about our beliefs. This has the important implication that the pessimist's argument must have two parts. For what he gets out of the induction over history, if he succeeds, is that we are likely often wrong in our theories, that is, our beliefs are unreliable. Recall that the position I am defending is not a general one about contemporary theories—that the successful ones are not too often wrong—but only the scientist's right to go on apportioning confidence in particular claims about unobservables, say about the interior of the Sun, to the strength of the usual kind of evidence she has for them, in the way a good scientist regularly does.

Unreliability is a property of *beliefs*, not of the convection currents in the interior of the Sun, and this presents an obstacle to the PI that has not been appreciated. Why should learning about our beliefs have an effect on what we think about the interior of the Sun? Facts about our beliefs are just not about the Sun, so how are they relevant? Put differently, what we believe about the convection currents in the Sun does not make a difference to what the interior of the Sun is doing, which corresponds to the fact that its apparent correlational relevance is screened off from claims about the Sun by other claims about the Sun. In gaining relevance of the past to the present by ascending to the meta-level, the pessimist has put in question the past's relevance to the confidences in particular theories that I am defending.

Any PI needs a justification not only for the inference from our predecessors' unreliability to ours (the horizontal inference), but also for the inference from our unreliability—being often wrong—to withdrawal of confidence in particular claims (the vertical inference). Having ascended to the meta-level to find premises that would not be irrelevant because of their difference in content, he must now descend if he is to deliver conclusions about our different content. Why should we think that an assumed general unreliability shows up as a falsehood here about muons, or there about quarks? How does any such inference go, and how is it justified?

I think second order beliefs do impose obligations at the first order, but why, and how it goes, are non-trivial questions. Elsewhere I defend a general answer to these questions, and so let pessimism live for another round. (Roush 2009, and ms.) The answer is that it is good to be calibrated, that is, for your confidence in proposition q to not only be appropriate to your evidence for q—say that it will rain tomorrow—but also to match your reliability in q-type questions—whether it will rain on day x—where reliability is an objective general relation between your believing q-like things and their being true. The paradigm case of calibration questions concerns a weather forecaster, whose reliability can be evaluated by track record, so the PI is well-suited to take advantage of this notion.

What is relevant here is that the rule I have defined (Roush 2009) demands a proportionality that explains why if we have evidence that the history of scientific failures is sufficiently relevant to our general reliability about things like the existence of muons and quarks, then our scientist does have a problem with particular claims about muons and quarks. This is because if a high fraction of our predecessor's theories about q-matters have been wrong, say 80 %, then the fraction of q-matters we are likely wrong about is 80 %, and only 20 % are still to be considered right. The calibration norm then says to dial down the confidence in any such particular claim to 20 %. This explains nicely part of the intuitiveness of the pessimistic argument, and shows that the vertical inference is defensible.[8] *If* the pessimist gives us reason to believe we are unreliable in q-like matters, then we should dial down our confidence in q.

6.5   Cross-Induction on Method

The horizontal inference from the past to the present is where the pessimist's stumbling block lies, in the question of whether our predecessors' unreliability is sufficiently relevant to our work to give us an induction to our own unreliability. For this, that in virtue of which we are justified must be sufficiently similar to the way our predecessors were justified. Why should we think this, and why do people actually think this? It is true that we are all doing induction in the broad sense, but we can say a little more and it has another name. We all, after all, used the Scientific

---

[8] Unfortunately for the pessimist, I have found almost no philosophers who agree with me about calibration and the role of re-calibration in assimilating evidence about ourselves, and despite a lively discussion of higher-order evidence going on in philosophy, no one has offered an alternative general account of how we are to take higher-order evidence into account or what justifies that. I maintain that the pessimist needs an account here, but since I believe there is one I grant him what he needs.

Method.[9] Even though we avoid the language of scientific method, the rationality-philosopher's search for the most general rules of inductive reasoning does implicitly keep a focus on one method, in aiming for the minimum number of principles from which we could derive all of the various more particular rules we see as being followed. And unabashed reference to the Scientific Method goes on as ever among scientists and laypeople. This focus on general method is a strong force I think in the grip the PI has on philosophers and others.

At least note this: since method is how we get from sensory irritations to beliefs general enough to be the conclusions of scientists,[10] the PI is maximally powerful when past and present scientists all use exactly the same method. Thus, we could blunt the PI by denying that there is any shared method at all. This is not an option for those of us who think there is, in the sense of basic forms of inductive inference and the demand for probabilistic coherence, and even some more sophisticated principles of evidence management. However, even granting the existence of shared method, or rules, or generalizations about sanctioned belief behavior, it is a mistake to think that that is enough to make the pessimistic induction work here.

This is because cross-inductions are available. A cross-induction does not require that the two populations be different in *every* way, but only in some way plausibly relevant to the projected property. And there are many more specific things to say about methods that are relevant to the effectiveness of our belief-forming practices at giving us true theories, that is, to our reliability. Those specific things are different for different contexts and questions, and even for the same subject matter, methods are different *between our predecessors and ourselves*. There is a lot more method than general philosophers of science tend to think about. Any procedure, tool, experimental design, protocol, instrument, is a method, because it has generality. It is repeated, held the same over cases of probing the world, and something has to be because we need sufficient sample sizes of evidence produced and evaluated in the same way to do legitimate positive inductions about particular matters. And though there is a great deal of retention of method at the general level, even there there are always new statistical methods, procedures, distinctions, and tools that are added to the methods we retain from past science. Statistics is itself a science, and it expands over time.[11] Thus, we will escape the PI on those particular occasions when cross-inductions on method are available, and they very often are when we compare ourselves to much of the history of science. Under-description of scientific method

---

[9] We tend to be reluctant to extend this compliment to science before the modern era, but I think a pessimist believes there is enough failure to get true theories in the modern era to allow the pessimistic argument to proceed. Note that it is not necessary that one associate the similarity of justifiedness the pessimist attributes to us and our predecessors with scientific method, in order for my cross-induction below to succeed against it.

[10] One might think the similarity between our predecessors and ourselves is that we all have inferred truth from empirical success. However, that is a method because it is a rule that we use to justify a move from results to conclusions. It is also an exceedingly general method, and more particular things there always are to say about methods can cross this method too in the way discussed below. For example, there are ever more and different ways of evaluating empirical success.

[11] See Glymour 2004 for some examples of this.

conceals irrelevance of the premise (here data about past scientists' justifiedness and failures) to the conclusion (here the claim that we are unreliable).

The specific, concrete differences of such belief-forming methods that you will find a good scientist counting as rendering previous failures irrelevant are differences that her evidence and background knowledge say are plausibly positively relevant to reliability. For a particular case, suppose your predecessor using chlorine in an experiment failed to get the expected answer, failed, in particular, to detect neutrinos at all the energy levels expected. Suppose you want to do the same experiment using gallium. A PI could say that since that same experiment failed enough times before, you are not justified in conducting it again. In this actual case, it was the same experiment to a significant degree, so there is a similarity base. But obviously, if a good scientist is proposing to do that experiment again with gallium that will be because he has reason to believe that the material—chlorine vs. gallium—could well make a difference to the results, good enough reason to make it possible to secure large amounts of funding for the quite elaborate operation. The failure using chlorine gives you good reason to believe the experiment will fail with gallium only if you do not have reason to believe the difference in material could make a positive difference to whether you detect neutrinos that might be there.

The pessimist is right this far: if you do exactly the same experiment a thousand times then you should not expect a different result the 1001st time. Ten times would probably be sufficient. But though I am no expert in history, I am confident that the very same experiment over and over, with different personnel and freshly laundered lab coats but no changes plausibly relevant to reliability, is not what the history of science looks like.

Of course, everything is different from everything else. For every single experiment there will be some respect in which it is different from every other. Why does this not imply the ridiculous conclusion that all past failures of science are irrelevant to whether our work is reliable, and are legitimately ignored? Part of the answer is that the difference that crosses the induction to our likely failure has to be something we have reason to believe is relevant to the property the pessimist is projecting and the scientist is crossing, here unreliability. There is a fact of the matter, and often evidence about, whether a difference is relevant to reliability on question q, and many are not. For example, often the experimenter wears a different shirt when he runs the same experiment on a different day, but typically we do not think that will affect the results. If so, then yesterday's failure is not irrelevant just because of the different shirt. The fact that you use your method today or tomorrow usually is not relevant, although it will be if you are studying astronomical events like eclipses. You may have good reason to believe that whether you do it in Chicago or New York will not, per se, be relevant to the outcome, in which case you cannot ignore a failure the same experiment had in the other location just because it was a different location. Generally, which earrings I wear will not matter, but if they are made of heavy metal they might interfere with a magnet, so taking them off could make a difference, and if they are big and bright enough they might be distracting in a cognitive experiment with babies. Some properties are relevant to reliability on question q, and some are not. There is a fact of the matter that depends on the case, and that scientists make arguments about.

Whether the PI over the history of science has any doubt to contribute with respect to a particular hypothesis comes down to the question whether the method used to investigate that hypothesis is different from methods used in all past failures in investigating hypotheses, in a way sufficiently relevant to reliability on the scientist's current subject. As we have just seen, this is a type of question the good scientist addresses explicitly as a matter of course in investigating the hypothesis. The question whether using gallium or chlorine is likely to make a difference to the results will be discussed thoroughly in any grant application for the type of solar neutrino experiment mentioned. So, the scientist addresses the PI over the history of science in doing the science itself. It follows that if the pessimist is going to give the scientist reason to doubt her hypothesis, he is going to have to argue not about theories and reliability in general, but about gallium, and whether its differences from chlorine are relevant to the energies of the neutrinos that can be detected with the given apparatus, and that will be a discussion with the scientist.

The pessimist might object that our scientist does not consider all of the cases included in the PI merely by a discussion of gallium. However recall that the question is whether this scientist's method is relevantly similar to those used by the scientists of the past who failed. His method is similar to those of Priestley with phlogiston and Lamarck with spontaneous generation and inheritance of acquired characters in roughly the way that paper towels are similar to swans. The relevant differences are so obvious that scientific journals economize on space by not requiring discussion of such comparisons. The scientist need not have explicitly considered them in order to take such cases properly into account.

One might suspect that I am begging the question, appealing to science to justify science. Surely the scientist is only justified in ignoring the general pessimistic induction if she is *justified* in thinking that the difference between chlorine and gallium, and that between the neutrino detector and a microscope, and between the neutrino detector and a bell jar, etc., are relevant to the reliability of detecting neutrinos. What right have we or they to think that? Moreover, why is my demand that the pessimist engage the scientist at the object level not gratuitously ignoring the issue?

I do not need to know whether the difference between chlorine and gallium and that between a neutrino detector and a microscope are relevant to reliability in detecting neutrinos, nor do the scientists' arguments about these matters need to be successful in order for my deflection of the PI to succeed. My point has been that these are the kinds of questions the pessimist's argument depends on and to which he needs negative answers if he is to succeed. They are questions about which particular differences of method are relevant to reliability of particular conclusions. They are particular because the scientist needs only one relevant difference of method in order to have a cross-induction against the relevance of a past failure. Every particular feature of a method in a particular case of our science is thus a potential threat to the pessimistic induction. All such features taken together exhaust the potential doubt a PI over the past could muster. It happens that every day scientists address, or are ready to address, particular claims about what methodical differences are relevant to the reliability of their particular results. Since every such feature is a threat to the pessimist's argument, his argument can only succeed if he takes the fight to the scientists themselves, and argues, for example, that the neutrino detector is not different from a microscope in a way that is plausibly relevant to reliability at detecting neutrinos. Perhaps the pessimist would succeed, but it would not be via an induction over history.

My claim is not that PIs *never* work, and therefore not that *all* past failures of scientists to get true theories are irrelevant to our confidences in our theories. Millions of PIs are good, and we do most of them implicitly, often without blinking. Some PIs—like the one from the whole history of science to a particular current hypothesis—are bad. To be a good scientist requires addressing, or being ready to address, questions of which particular similarities and differences to previous efforts are relevant to the reliability of one's particular results, and in doing so one addresses all of the doubts the history of scientific failures has to offer.

6.6   Rationality and History

Thus the specific differences and discontinuities of method over historical time—the things that historians are especially interested in—positively support, and are indeed essential to, the rationality of scientists' beliefs. It is useful to compare this situation with the difficulties that came for the rationality of science when Kuhn said there were global discontinuities called paradigm shifts that affect virtually everything about the way that a science operates—assumptions about the basic building blocks of the world, what are meaningful questions and sensible ways of going about answering them, and so on. The main, and familiar, problem about rationality is that if everything changes at once then there is nothing unchanged through a paradigm shift that could be a neutral arbiter between the before- and after- theories, to tell us why the change is rational or justified. The arbiter used to be observations, but these are theory-laden; what you see depends on which theory you already subscribe to. I will call this the *neutral-arbiter problem*. Obviously what I have said does not address this problem, but what I have argued has similarities to, and differences from, what Peter Galison said to this problem.

Galison points out that it is not true that all of science changes as a block (Galison 1997, pp. 701–844). Go to a higher resolution and you will see that there are more levels than theory and observation. There is not only experiment testing theory, but material culture and computational methods among many other things, and the different cultures have quasi-independent inner logics driving them. So, layers typically change at different times according to their own needs and objectives, which are not always that of testing high theories. The continuity at one level can give you a vantage point from which to judge the wisdom of changes in another. And it does not have to be the observation level that is always the unchanged neutral party. Thus the intercalation of these layers is part of the epistemic strength of science.

Perhaps there is a non-trivial level of description at which everything changes at once, but for many purposes it is an under-description, and the under-description matters to rationality because it hides neutral arbiters. The move I made above has in common with Galison that you find your way out of skepticism by taking more specific facts into account, and explaining why they matter to rationality. However, Galison's argument rescues the epistemic strength of science by finding continuity over time despite the temporal discontinuities. In the rope metaphor that he uses, the existence of quasi-independent strands is crucial to a rope's strength because when one strand is strained, the others are not breaking. I am addressing a different problem, which is a dual of the neutral arbiter problem because in the PI it is the continuity and similarity of method over time that appeared to create a problem. And the rationality of typical science is assured against this problem, I think, not despite differences over time but because of them. Earlier I pointed out that identical method in every instance of science would produce the most powerful possible pessimistic induction. The dual of that here is that a radical paradigm shift in which everything, even the more specific layers, really did change at once would be the ultimate weapon against the PI. It would prevent any induction from previous science to our own, whether negative or positive, because if everything were different there would be no similarity base at all. Of course, if we claimed a paradigm shift in that radical sense then an answer to the PI would come at the expense of an answer to the neutral arbiter problem.

However, just as Galison did not need to claim that all of the layers of science remain the same over time in order to address the neutral arbiter problem, so, too, no claim of difference all the way down was necessary for my defense of the typical work-a-day scientist given above, since crossing a PI does not require that the method of investigating a hypothesis be different in *every* way from the method that tried and failed before. It only needs to be different in some way that we have reason to believe is relevant to reliability. There is always a pool of similar episodes that are relevant to whether a scientist should trust the results of what she is doing now, and she must—and I say typically does—take their failures into account. However, that pool gets smaller the more fully what she is doing now is described. Consistently with Galison's view, both similarity and difference, continuity and discontinuity, are actual and necessary to the rationality of scientists' beliefs.

6.7   Application

I will illustrate how this argument goes for a recent version of the PI, before going on to reply to objections. In Kyle Stanford's PI, the similarity base is not explicitly method, but the induction can be crossed by means of facts about method. His argument uses as a similarity base the fact that we and our predecessors are subject to unconceived but conceivable alternative hypotheses about unobservables that are equally compatible with, and explain, our evidence (Stanford 2006). We know that our predecessors were subject to this because we have since conceived relevant alternatives they did not. There is no reason to think we are different in this, so we can expect our successors to conceive of alternative explanations for our evidence that we have not. Our predecessors were often wrong about unobservables[12] and this was connected to their failure to conceive of conceivable possibilities. We can therefore also expect to be shown wrong in our hypotheses about unobservables because of our similarity to our predecessors.

I do think we can expect to be shown wrong, but the significance of that generic fact is questionable, as I will discuss below. The first flaw in Stanford's pessimistic argument is that it does not take into account changes in methods, specifically methods for ruling out alternative hypotheses about unobservables. The kinds of examples Stanford deals with are hypotheses about the mechanisms of heredity and they illustrate nicely how limited our intuitive imagination is. But we have statistical methods for ruling out alternative hypotheses that do not require intuitively imagining the mechanisms or objects that could be involved (Roush 2005, pp. 218–221; 2010; Glymour 2004). Unconceived does not imply not ruled out. Thus, past scientists lacked many methods that we have for ruling out unconceived alternatives, and ruling these out was presumed by the pessimist to be relevant to reliability. Induction crossed.

Since we never have full evidence, we cannot suppose there is ever a stage at which we have ruled out all unconceived conceivable alternative possibilities, even if we employ different methods every time. But this remaining similarity between our predecessors and ourselves is not as significant as many suppose. That there exists at least one alternative explanation of one's evidence means that it is possible one is wrong, but says nothing about how plausible that is, and thus nothing about the degree or extent of our unreliability. One might respond that though the mere existence of such hypotheses at every stage is not a problem the number of them surely is. But for this gambit the pessimist will need to argue that the number of remaining alternatives is always high, and I do not see how he knows that. It is common even to suggest that the sea of remaining conceivable hypotheses that give explanations of our evidence will be infinite no matter what we do. But the evidence of history used by Stanford gives no argument for these claims, since we have (intuitively) conceived of only finitely many possibilities that our predecessors did not, and a small number at that.

If we grant the infinity of that set of unconceived conceivables for the sake of argument, the idea that it is a problem is based on some misconceptions. One source of the response is the fact that any finite number divided by infinity is zero. Thus, ruling out any finite number of further alternative hypotheses does not constitute progress on the alternative hypotheses problem because it does not raise the fraction we have ruled out and so does not raise the probability of our original hypothesis. However, while nineteenth century scientists, even physicists, ruled out hypotheses seriatim, modern statistical methods allow us to rule out classes containing an infinite number

---

[12] Note the dependence of this premise on a high base rate for our predecessors' failures. This is necessary to come to a conclusion that we are *likely* to be wrong, and, as mentioned earlier, has been argued to be impossible to assign (Lewis 2001; Magnus and Callendar 2004).

of unconceived alternative hypotheses in one stroke (Roush 2005, pp. 218–221). And even supposing that there remain an infinite number of alternatives there can be a clear probabilistic sense in which the proportion remaining has been decreased. Mathematically, this only requires that all hypotheses are assigned finite non-zero weights that sum to one, which can be done using any convergent infinite series of fractions that sums to one.

We can rule out possibilities without conceiving of them, an infinite number at a time, but if we must always suppose some possibilities remain, this suggests another intuitive problem. We will never get to the end of this space, let us suppose, and so it seems we cannot span it or take its measure. Thus, how can we ever legitimately estimate how far we have gotten? This is a vague thought that corresponds to two real questions, but to neither of them are failures in the history of science more relevant than scientists already take them to be. The first question is how our scientist spans that space to come up with a particular probability for a hypothesis about the convection currents of the Sun. But objecting to that will require arguing with him either about the details of his evidence for that particular hypothesis or about his estimation methods, and will not require or be helped by a PI over history. The other possible argument that could be attempted here is a general one: namely, that scientists cannot possibly have grounds for evaluating the catch-all term[13] and thus not the probability of the hypothesis itself. That would be a conceptual matter to take up with a statistician, or a statistically inclined philosopher, or the scientist so inclined herself. The point is that it would not require, or be helped per se, by reference to the general fact that scientists have failed in the past.

The pessimist might protest that it is unobservable claims that are at issue, that history shows there is something especially recalcitrant about them. Many successful theories were wrong in their generalizations about observables too, but I will put that aside. This distinction makes no difference to my argument because I do not have to show that method changes make a difference to reliability about unobservable matters, even in particular cases. The unobservability of neutrinos in the example above made no difference to the fact that the issue of whether scientists are justified in believing things that they do about neutrinos depends on whether there is a cross-induction via the method used to establish things about neutrinos, which depends entirely on things like whether changing the material to gallium plausibly makes a difference to the result. This is an issue, and a kind of issue, that scientists address explicitly, so it is the scientist whom the pessimist must challenge.

Must, that is, unless the pessimist can make an argument that it is, in principle, impossible for method differences to change one's success at claims about unobservables. This might be argued on general empiricist grounds, though I think that is unsuccessful (Roush 2005, Chap. 6). But those arguments are not a PI, and if successful would not need a PI. However, perhaps a new PI could be made about method itself. We see that we have relevantly different methods from our predecessors for going at claims about unobservables, but they had apparently relevantly

---

[13] This is the probability of your evidence given all logically possible alternatives to your hypothesis.

different methods from their predecessors too, and little good it did them for they still came up with theories that are false if our theories are true.[14] This PI is aimed at all of the factual claims—that this or that method change *is* plausibly relevant to reliability—that a cross-induction could rest on.

However, this gambit is untenable. For what justified the pessimist's doing an induction here rather than a counter-induction? A counter-induction would have taken us to the conclusion that having failed in the past, this time the new and apparently more reliable methods *are* relevant to reliability about unobservables. Since the argument's conclusion is about unobservables, in particular about the relevance of method changes to reliability about them, the inference that was made is justified only if we think induction is a more reliable method than counter-induction at getting true conclusions about unobservables. But then this argument's conclusion that no method is relevant to reliability about unobservables undermines its own justifying inference rule. The argument's conclusion prevents legitimate inference to that conclusion. Another caution is in order with this conclusion, of course, since we could have gotten to it by Hume's argument that no inference or method makes it more rational to believe this versus that about the unobserved or the unobservable. Thus, if we want to establish that conclusion then reference to history is an unnecessary detour.

6.8   Too Good to be True? The Size of Potential Error

By now it may seem that my conclusion is just too good to be true.[15] Intuitively it seems that there must be something right about the PI. There is something right, though as I will argue it too is already taken into account in good scientists' particular judgments. That there is something right comes from the fact that induction is not deduction. Just as inductive support comes in degrees, so does every cross-induction come with a degree. There is a degree to which you are justified in believing the cross-induction property is present, and a degree to which you are justified in believing it is relevant to your reliability on q, those two combining for a cross of a certain

---

[14] Thanks to Bill Talbott for this argument. Note once again the dependence of its premise on a high base rate of falsehood of past successful theories.

[15] Thanks to Catherine Z. Elgin for the very helpful objection addressed in this section. There is another apparent way of arguing that a history of past failures has got to make a difference that our particular judgments do not already pick up on, which is imagining the same current situation of evidence and theories but preceded in one case by a history of failures and in the other case by a history of successes. Surely that makes a difference to the confidence we are entitled to have in our theories (Thanks to Shelly Kagan for this objection.). The problem with this argument is its premise. If our theories have all successes, then there cannot have ever been evidence inconsistent with our theories, since it would be part of our evidence pool. So if the same amount of evidence was collected in the two pasts then there is much more positive or neutral evidence for our theories than there would be with a history of past failures. That means that it is not possible to have the same current theory-evidence situation with the two different histories.

degree. Some relevance of the past to the present remains because there are still similarities between our predecessors and ourselves. The fact that we are all human beings doing the scientific method in the most abstract sense is the scaffolding for the thousands of more particular features of our methods, so its relevance is not zero. Thus, something of the general PI remains, which raises the question how strong it is. It is easy to see from what I have argued above that the degree of legitimate PIs and crossings by novelty of method will co-vary with the degree of similarity and dissimilarity of method. I will now argue that both of these co-vary with logical strength of hypotheses. This means that to be sensitive to the logical strength of hypotheses is already to be sensitive to the degree to which pessimistic inductions work.

First, it is surprisingly rarely appreciated that the admission that a theory is very likely false is thoroughly compatible with being highly confident of each of its particular claims that it is true. At least it is rarely appreciated that this is a good thing.[16] To illustrate, consider the example of the Standard Model of particle physics combined with auxiliary assumptions in comparison to a particular claim that follows from them, such as the existence of the Higgs boson. On the basis of successful experiments, physicists may be, and some are, quite confident though not certain that the Higgs boson exists, while also being confident that the Standard Model is false. The rationality of this can be seen if we represent the rational confidence of a subject as a probability. Then we would formulate the claim that the subject's degree of belief in q is x as $P(q) = x$, the probability of q is x. If so, then the fact that a scientists' degree of belief, x, is 99 % instead of 100 %—confident but not certain—makes a very big difference to what probability rationally requires when a single claim like the existence of the Higgs boson is conjoined with many others.

If one is certain in a claim, then probabilistically one cannot coherently revise it, meaning that one regards it as impossible that the claim is mistaken. It also requires that one be certain of its conjunction with other claims one is certain of. In contrast, if one has even a sliver of a doubt about individual claims, and those claims are independent, then one's confidence in the conjunction must be exponentially lower than that in any of the individual claims. If there are 16, let us say independent, claims of existence of elementary particles, then even if the scientists are extremely confident about each, say 99 %, rationality requires them to have a 15 % confidence that at least one of them is wrong. This doubt in the conjunction grows with the number of conjuncts: with 40 independent claims, the required degree of doubt in the conjunction is 27 %. The degree of required doubt increases even faster the lower the confidence in the original individual hypotheses. If one is 95 % confident in each of the 16 claims about the particles—only 4 % lower than just supposed—then one will be required to be *more confident than not* that at least one of them is wrong: 57 %. (Starting with 99 % it takes 100 claims to get to more likely than not that the conjunction is false.)

---

[16] An exception is Philip Kitcher (2001). Many take this probabilistic fact to indicate a "paradox of the preface," mistakenly in my opinion (Roush 2010).

To see what this tells us about theories, we have to say what a theory is. If we idealize, then a theory can be written as a set of a few independent law-like generalizations. However, that will have no empirical consequences for this world, the actual world, without adding a lot of auxiliary assumptions about this world. For example, Newton's theory of mechanics can be compactly expressed as three laws, but one must specify where the massive objects actually are at a given time, how massive they are, and so on, in order to figure out what this theory says about where they are going to be at a different time. If seen as a proposition, then a universal, substantive theory is a huge conjunction, and with the number of claims increasing at this level, the required confidence that the conjunction is wrong increases dramatically. If we have a million claims, then even if we have 95 % confidence in each we are required to be more sure that at least one of them is false than we are sure of any individual one of them. In our example, if we are 95 % confident that the Higgs boson exists, then to remain rational we must be 96 % confident that the Standard Model is false. From the other side, even if we are 96 % confident that the Standard Model is false, very high confidence in the existence of the Higgs boson does not make us irrational.

One might object that the magnitude of this admission that high theories are likely false is inadequate to address the point of the pessimist. This is merely confidence that at least one of the million claims is false. What we see in the historical record is cases where a big part of the big idea was wrong in a big way. Should logic alone make us confident of the same about our own theories, and if so, then would that not amount to winning the battle but losing the war against the pessimist? One might mean several different things by "big", but we can address the objection by measuring size of error as how many or what fraction of one's claims were wrong.[17] This would capture, for example, the idea that a big claim has many implications. Consider a theory with 1 million independent claims as earlier, and suppose that we are 85 % sure of each of them. Then we will be obligated to be 97 % confident that the theory is false somewhere. But also, at this 85 % level it only takes five claims to be required to be 56 % sure that one of them is false.[18] That, as like as not, one of every five of my claims is false is a substantial admission: one fifth of 1 million is two hundred thousand. Yet this does not prevent the rationality of my 85 % confidence in each one of the 1 million claims. This is as it should be because the big admission that two hundred thousand of my claims are more likely than not to be false, and the reference to the "big" mistakes of history, give us no hint of which of my claims are at fault. Recall that it is scientists' right to go on with their practice of making this or that particular claim in keeping with their usual evidence and arguments that

---

[17] Is the structure of a theory the big part, or is it the types of entities the theory takes to exist that are big? Is the hypothesis that the ether exists big because the ether was supposed to cover the entire universe and affect every motion? Or was it small because it was a single proposition that, as it turned out, is independent of much of the rest of the theory it was housed in? Answering these questions would do much to carve out one's brand of realism or anti-realism, which is not my purpose here.

[18] If my confidence in each is 0.75 then I must be 0.76 confident that one of every five claims I made is false.

I am concerned to defend, and our example of a big admission does not impose a confidence drop below 85 % on any particular claim. A general theory is equivalent to a huge conjunction, so we should be well aware without looking at our predecessors that it is very likely to be false, and it is clear on reflection that that does not impose a low confidence for any particular claim.

Logical strength of a hypothesis also makes it more susceptible to a PI by making it harder to rescue by a cross-induction on method. Consider the 16 particles of the Standard Model. Verifying all of those particles supports the theory to some degree, and the whole set of these verifications gives stronger support to the theory than any subset of them would. However, these particles were verified using a very wide array of types of methods, i.e., particle detectors. The bubble chamber is different from the spark chamber, and neither is much like the Large Hadron Collider. To defend the Standard Model against a PI, we should appeal to differences between the method we used to test it and those our predecessors used on their theories. But if we were to state the method by which this theory was tested or supported, it would have to be a quite generic saying because it would have to be true of all of those methods that went into the verification, and relatively little is. There will be some statistical methods that all of our particle experimenters have used and their predecessors did not have, but what is common in the verifications of the particles does not go a great deal beyond that. By contrast, someone who used the bubble chamber to detect kaon decay would have a great deal to say about how his method was relevantly different from those of his predecessors who tried to detect the unseen. We necessarily have less material for a cross-induction on method on behalf of a theory than we will on behalf of particular, logically weaker, claims that follow from the theory. Logically weaker claims are not only easier to support by evidence (other things equal), but also more resistant to the PI. Notice once again that the distinction between observable and unobservable made no difference to the arguments of this section. Even supposing it exists, the Higgs boson is unobservable.

6.9   Conclusion

The history of science appears to pose two threats to the rationality of science, one due to radical paradigm shifts, the other due to our predecessors' track record of failures to get true theories. These challenges are duals in that one rests on the consequences of too much discontinuity, the other on those of too much continuity. However, these apparent problems are illusions due to under-description, and they disappear when with more specific descriptions of the history we find additional continuity and discontinuity respectively. It appeared that history cast doubt on the rationality of science, but the rationality of science is saved by an eye for detail that is characteristic of the historian.

In particular, the pessimistic induction over the history of science is powerless to create justified doubt about our particular hypotheses that is not already addressed, or prepared for, in good scientists' arguments about particular conclusions. For scientists to become doubtful about particular hypotheses on the basis of a general induction over history would be for them to double-count the evidence, and to ignore the relevance that difference of method has to reliability. It would be as if, like characters in an Ionesco play, they were to infer that paper towels are white from seeing white swans, because they saw them in the same country.

Acknowledgment I am grateful to many people for comments on this material over the years. They include Bill Talbott, Andrea Woody, Arthur Fine, Catherine Z. Elgin, Shivaram Lingamneni, Shelly Kagan, George Bealer, Michael Della Rocca, Zoltan Szabo, Edward Irwin, Peter Lewis, Joseph Carter Moore, and Harvey Siegel. I would like to thank Peter Galison, Simon Schaffer, and Dave Kaiser for everything they have taught me.

References

Donovan, A., L. Laudan, and R. Laudan. 1992. *Scrutinizing science: Empirical studies of scientific change*. Boston: Kluwer.

Galison, P. 1997. *Image and logic: A material culture of microphysics*. Chicago: University of Chicago.

Glymour, C. 2004. The automation of discovery. Daedalus Winter: 69–77.

Harman, G. 1980. Reasoning and evidence one does not possess. *Midwest Studies in Philosophy* 5 (1): 163–182.

Hollinger, D. 1973. T.S. Kuhn's theory of science and its implications for history. *American Historical Review* 78 (2): 370–393.

Kitcher, P. 2001. Real realism: The galilean strategy. *Philosophical Review* 110 (2): 151–197.

Laudan, L. 1981. A confutation of convergent realism. *Philosophy of Science* 48:19–49.

Lewis, P. 2001. Why the pessimistic induction is a fallacy. *Synthese* 129:371–380.

Magnus, P. D., and C. Callender. 2004. Realist ennui and the base rate fallacy. *Philosophy of Science* 71:320–338.

Poincare, H. 1905. *Science and hypothesis*. London: Scott.

Roush, S. 2005. *Tracking truth: Knowledge, evidence, and science*. Oxford: Oxford.

Roush, S. 2009. Second-guessing: A self-help manual. *Episteme* 6.3:251–268.

Roush, S. 2010. Optimism about the pessimistic induction. In *New waves in philosophy of science*, ed. P. D. Magnus and B. Jacob. London: Palgrave-MacMillan.

Stanford, P. K. 2006. *Exceeding our grasp: Science, history, and the problem of unconceived alternatives*. New York: Oxford.

Chapter 7
What do Scientists and Engineers Do All Day? On the Structure of Scientific Normalcy

Cyrus C. M. Mody

When I entered Cornell's Ph.D. program in Science and Technology Studies in 1997, *Structure of Scientific Revolutions* was naturally one of the first books I was assigned. I came into the program with a bachelor of arts in engineering, so perhaps I was predisposed to appreciate Kuhn's lively mélange of the humanistic and the technical. Kuhn did me a bit of a disservice, though, in that his writing was so clear, readable, and relevant to both what I knew and what I wanted to know more about that he set the bar very high for anything else I would read (or write!). Certainly, few books since then have matched Kuhn in setting forth a bold, intelligible argument in such an engrossing way. Still, I look back on that first reading of Kuhn as a sign that I made the right choice to be in a field where I work alongside other scholars who also look back at Kuhn as a common point of reference.

Curiously, it was the early chapters of *Structure* on normal science that really spoke to me and that continue to inform my work. I think it's clear that Kuhn's own ambitions had more to do with the later sections on revolutionary science, and certainly those are the ones that have generated the most heat. But, back in the late 1990s, to this lapsed engineer and aspiring laboratory ethnographer, the differing logics of normal science across time, culture, and discipline seemed a much more fruitful line of investigation than the endless and acrimonious debates about incommensurability, relativism, and the uncertain reality of reality (e.g., Koertge 1998; Labinger and Collins 2001).

Not, of course, that the line of work that spun off from Kuhn's thoughts on revolutionary science hasn't been extraordinarily generative. Particularly influential for me in that regard has been the long, rich tradition of so-called "controversy studies" associated with the "strong" Edinburgh school version of the Sociology of Scientific Knowledge (e.g., Mackenzie 1990; Shapin 1975; Edge and Mulkay 1976) as well as the so-called Bath and York schools (Collins and Pinch 1982; Pinch 1986; Ashmore et al. 1989). Even in these studies of potential paradigm shifts, however,

C. C. M. Mody (✉)
Department of History, Rice University, Houston, TX, USA
e-mail: Cyrus.Mody@rice.edu

© Springer International Publishing Switzerland 2015
W. J. Devlin, A. Bokulich (eds.), *Kuhn's Structure of Scientific Revolutions—50 Years On,*
Boston Studies in the Philosophy and History of Science 311,
DOI 10.1007/978-3-319-13383-6_7

the persuasive force and many of the lessons of the SSK controversy studies depends to a great extent on a close examination of Kuhnian normal science. That's in part because virtually none of the contemporary controversy studies were able to follow a case of a successful overthrow of some set of major, established scientific facts. Kuhn's point that revolutions happen so rarely that few scientists live through a paradigm change was seemingly confirmed. Indeed, in order to catch a revolution in progress, sociologists had to resort to historical controversy studies, such as Steven Shapin and Simon Schaffer's *Leviathan and the Air-Pump* (1985).

More subtly, my early reading of the SSK controversy studies suggested, to me at least, that the whole argument of this genre depended on the discovery that the seeds of uncertainty, controversy, ineffability—seeds that sprout during times of revolution—are present but relatively unproblematic during times of normalcy.[1] Take, for instance, Harry Collins' (1985) *Changing Order*, a three-part study of applied scientists building a laser, astrophysicists building and debating gravitational radiation flux detectors, and parapsychologists trying to communicate with plants. Like Kuhn, Collins is quite obviously more interested in revolutionary than normal science—hence the *changing* order of the title, as well as the subsequent four decades and several thousand published pages of text in which Collins has stuck with his gravitational radiation researchers in hopes that they will unearth something paradigm-changing.

Collins' argument about gravitational radiation research is that it is such a complex endeavor that its practitioners are unable to describe all of its intricacies even to each other. Even if they could, he claims, there is a great deal of tacit knowledge embedded in gravitational radiation flux detectors that is beyond the conscious grasp of even those who possess it, and which is therefore only contingently available to inflect debates about whether a particular detector is working properly and has or has not detected the passage of a large gravitational wave. That argument has been contentious, and even Collins has retreated from his less cautious formulations of it (Collins and Evans 2002).

Yet the normal-science preamble to that part of *Changing Order's* argument seems relatively unexceptionable—Collins follows a group of applied physicists as they try to build a TEA-laser in the early days of that technology. They fail repeatedly, even though one of them has built such a laser before and has all the formal knowledge needed to do so again. Eventually, *something* gets resolved and the laser begins to vaporize concrete—the rather unambiguous measure of whether it is working or not. Yet even though they eventually succeed in a task that, while difficult, is nowhere near the complexity of a gravitational radiation flux detector, the reason why they succeed remains a bit of a mystery. Collins hints, at any rate, that the scientists' explanation for why the laser suddenly started working is provisional and unimportant to them.

Indeed, it is not uncommon for scientists to volunteer skepticism of their own such explanations in interviews. Here's an example from an interview I conducted

---

[1] Thomas Nickles also stresses the similarity between normal and revolutionary science. See Nickles (2002).

with an early scanning tunneling microscopist describing how he overcame a similar experimental obstacle:

> We thought we eventually traced the problem to brass screws which were used in the sample holder. The brass screws contained zinc I believe and apparently the story was—I guess I still am not sure even now that this was what the problem had been—but the hypothesis was that this zinc ... in the screws was in a part of the sample holder which got very hot during the part of the cleaning procedure where you anneal the silicon. ... Zinc is volatile enough at those kinds of temperatures that you can get reasonable partial pressures in the chamber, maybe not a lot but enough that you could get a fraction of a monolayer of zinc on the silicon. Some of these heavy metals on silicon form silicides and are known to roughen the surface significantly. And we were seeing rough surfaces in which we couldn't see any particular atomic order.... So once we realized that was a possibility we replaced the screws, sent the chamber back and had it cleaned.... Here's why I don't know whether that's the real explanation, we also changed some other things in that procedure at about the same time and it started to work. So whether that did it or some of the other things we changed did it I'm not sure, and we didn't really care. That's not what we were researching. We just wanted it to work.

In other words, Kuhnian normal science is often less concerned with Truth and Knowledge than with getting things "to work." Scientists might be somewhat more focused on capital-T Truth and capital-K Knowing when embroiled in a controversy or a paradigm shift, but those are exactly the conditions in which everything is messier: the standards for what counts as "working" aren't settled, both the necessary tacit and formal knowledge are in shorter supply, and the variations among different groups' experiments are suddenly more salient than usual.

My point here isn't that Collins' observations about building a TEA laser confirm what he has to say about gravitational radiation research (or controversial, paradigm-threatening science more generally). Rather, I'm arguing that scientists' provisional insouciance regarding formal knowledge, and their willingness to endure large gaps in their understanding of how their experiments work, was an important but rather easily replicated and not terribly controversial discovery of SSK. Extrapolating from that discovery to bolder claims about leading-edge science might or might not be warranted, but, at the very least, SSK's textured view of wild-type normal science made problematic the more totalizing and rigid versions of some of mid-century philosophy's favorite hobbyhorses: demarcation, unity of science, scientific method, reductionism, the distinction between contexts of discovery and justification, etc. Those early SSK controversy studies didn't get everything right, obviously; but any contestation of SSK that ignores those studies' robust empirical findings about normal science—findings that, indeed, scientists themselves routinely echo—is not operating in good faith.

The other major Kuhn-descended genre of science studies that I was introduced to early in graduate school—in fact, the genre I entered graduate school to become a practitioner of—was the laboratory ethnography, as typified by a crop of studies of California labs in the late 1970s and early 1980s: Latour and Woolgar (1986); Knorr-Cetina (1981); Traweek (1988); Lynch (1985). Here, the connection to Kuhn's description of normal science is even more apparent. "Lab studies" are really just ethnographies of work. Their basic finding is that scientists' work habits look a lot like those of most other professionals, especially those in occupations that generate, manipulate, or disseminate information and/or that involve tinkering with materials and machines. Scientists spend much of their time writing (Latour and Woolgar 1986), gossiping (Garfinkel et al. 1981), promulgating-resisting-accommodating to bureaucratic rules (Gusterson 1996), venturing out to work sites (Latour 1999), etc. If one adopts the Martian perspective of an ethnographer, it isn't obvious just from their work practices which inhabitants of a lab are the "scientists" and which are the janitorial or cooking staff (Kelty 1997).

Does the quotidian nature of scientific work matter? Scientists, after all, also do science, which makes them special in the eyes of most late modern people—even if no one can cleanly demarcate science from other kinds of practice. At the least, though, lab studies' depiction of wild-type Kuhnian normal science gives ammunition to deflationary conceptions of science as, in Andrew Pickering's words, "practice and culture" (for such a framework, see Pickering 1995; for more examples, see the essays in Pickering 1992). Scientists make judgments in much the same way the rest of us do. Their judgment is considerably better than most regarding their particular patch of knowledge and practice, but that just means that we should extend to scientists the same degree of trust that we extend to experts in law, finance, education, insurance, management, etc. who also employ sophisticated bodies of "practice and culture."

Again, as with SSK, I am not arguing that laboratory ethnographers were correct in every extrapolation they made from their observations of wild-type Kuhnian normal science to bold claims about controversial or revolutionary science (or the validity of scientific knowledge in general). For instance, I'm sympathetic to (though not entirely persuaded by) Park Doing's (2009) insistence that laboratory ethnographers have not captured in real time a single instance of the "social construction" of knowledge. That may or may not be the case, and is certainly worth debating.

What's much harder to contest, though, are the baseline observations of ordinary scientific conduct that ethnographers have established. And from those fine-grained, easily-replicated observations, we now have a pretty good sense that, yes, scientists rely on all kinds of social cues to help them decide whom to trust, what data are robust, which results are important, which arguments are persuasive, etc. "Social cues" here means something like the list of rationale for belief or disbelief that Harry Collins (1985, p. 87) put together from interviews with gravitational radiation researchers:

- Faith in experimental capabilities and honesty, based on a previous working partnership.
- Personality and intelligence of experimenters.
- Reputation of running a huge lab.
- Whether the scientist worked in industry or academia.
- Previous history of failures.
- 'Inside information'.
- Style and presentation of results.
- Psychological approach to experiment.
- Size and prestige of university of origin.
- Integration into various scientific networks.
- Nationality.

Scientists don't, of course, *only* rely on such social cues, and their reliance on these cues is usually provisional. As I've argued elsewhere (Mody 2010), "contingent social cues are implicated in the current status of fact claims, in that they are part of a documentary method of interpretation. Scientists integrate what they know about their [and each other's] organizations and research communities into their understandings of technical measures."

And *vice versa*—they integrate their emergent understanding of instruments, theories, materials, experiments, etc. into their evaluations of each other. Again, from Mody (2010): "As the logjam of measures grows, scientists' understandings of their social worlds (who's competent, who's crazy, which disciplines are 'sloppy' or 'careful') shift, inextricably, with their understandings of nature (such that 'social' and 'natural' [or 'technical'] are entangled)." That shouldn't be a surprising observation, at least not to anyone who has read Kuhn and/or talked with practicing scientists.

Still, it's not an observation that's woven into many recipes for "the" scientific method or into prescriptions for the governance of science. Rather, most normative methodological and policy frameworks either ignore scientists' use of social cues or treat such practices as a particularly unfortunate consequence of the generally unfortunate fact that science is done by human beings. Indeed, the desire to ease humans out of science was an important factor behind the decades-long, and so far failed, attempt to turn artificial intelligence research into a branch of the philosophy of science (Roland and Shiman 2002; Dreyfus 1972, 1992; Collins 1990). Perhaps, ultimately, we will have machine intelligences that can do science—though, at the moment, it looks more likely that such machines will be extraordinarily sophisticated pattern recognizers (*á la* Siri or the movie recommenders used by Netflix and Amazon) rather than silicon philosophers. Still, that day looks much further away now than it did in, say, 1956.

In the meantime, my contention is that philosophers, historians, sociologists, and anthropologists of science could make a great deal of progress in understanding—and perhaps aiding—science by first acknowledging that scientists rely on social cues for plenty of good reasons. The humanity and sociality of science aren't perfect, of course, but they also aren't incidental or unfortunate features of the scientific enterprise. Rather, the lesson from Kuhn—perhaps his most important and robust lesson—is that the humanity and sociality of wild-type science are constitutive of the scientific enterprise as we know it. Human needs and desires, as promulgated through a variety of social formations, furnish the aims of science, the standards by which to recognize who counts as a scientist (and how good they are), the incentives for doing science, the paths to becoming a scientist, the means to do science, etc. Core features of scientific knowledge-making—such as, what counts as "objectivity" (Daston and Galison 2007)—are not set in stone, but have histories that vary over time and place, and are shaped by the aims of the societies that scientists are a part of.

It's easy to lose sight of that lesson, so long as Kuhn's contribution is taken to be about "the structure of scientific revolutions." Since that theme was the title of the book and the focus of Kuhn's passion, most people accepted that scientific revolutions were the ground on which debate would proceed. Accordingly, most of the conventional arguments over Kuhn's thesis (and those of his intellectual fellow travelers in science studies) have centered on questions of scientific change: Does science progress?; Do successive paradigms account for more of the world more accurately than their predecessors?; Can the basis for moving from one paradigm to another be (or be made to be) rational?; etc.

The question Kuhn started out with, however, turns those points of debate on their head. He wanted to know why the Ptolemaic system stuck around for as long as it did, despite flaws that were well known to its practitioners, and even well after viable alternatives had been put forward which addressed those flaws. Seen from that perspective, revolutions are merely Kuhn's stalking horse for exposing the structure of scientific normalcy. That is, if, historically, one can find a set of explanations (a paradigm, if you will) that we today believe was "better" than its predecessor and yet no revolution ensued to put the new paradigm into place, then that is a probe for understanding what the old paradigm achieved that the new one did not. It's also a probe for reflecting with more subtlety on what we mean when say that one paradigm is "better" than another.

Let me reiterate that point somewhat differently. My reading of Kuhn—as refracted through my training in the "Ithaca school" version of science studies—is that bodies of technical knowledge and practice (again, let's call them "paradigms") can achieve a certain obduracy despite the vast number of cogent objections that can be raised against them in part because scientists and engineers are able to make those paradigms workable, on a day to day basis, with respect to some ever-shifting set of aims promulgated relative to their professional communities and/or to various constituencies in the societies of which they are a part. That is, normal science keeps going, despite obvious anomalies and ignoration of open questions, because normal science achieves many more goals than just the clearing away of anomalies and open questions.

Thus, Kuhn offers a persuasive justification for putting to one side—or, at least, putting on the back burner—the question of how scientists and engineers ought to work, and instead gives us the grounds for asking how they actually do work. Normal science, despite all its flaws, is sustainable most of the time because it achieves something—not necessarily a more perfect picture of reality or an incontestable ontology or an ironclad method for generating new knowledge, but still, something that scientists and their patrons care about. So Kuhn allows us to ask—what is that something, and what does it tell us about science and its stakeholders?

In my own research and teaching, I often use a question and a very rough rule of thumb to try to identify what that something (or, usually, those somethings) might be for any given case. The question is the one in my title: what do scientists and engineers do all day? It's not, I think, a question that Kuhn himself would've asked, but it is a question that follows quite easily from his foregrounding of such mundane bits of scientific life as textbooks and problem sets. That is, Kuhn wasn't just interested in scientists' published works and polished statements, in which the messy tangle of quotidian demands has been scrubbed clean. Rather, Kuhn wanted to show that a paradigm informs and is emergent from every aspect of a scientist's life, even those aspects that the analyst might assume aren't germane to scientific knowledge.[2] Thus, Kuhn built his whole argument around the ordinary stuff of scientific practice—the stuff too common or too ephemeral to have garnered much attention before.

Those who followed him took that preoccupation with the ordinary, unfinished flotsam of scientific life even further. "What do scientists do all day?" is a question that can be read quite easily into most of the groundbreaking works of science studies of the past forty years. As I've indicated, it shines through in the laboratory ethnographies and controversy studies discussed above. It informs to a great extent the wonderful ethnographically-textured laboratory histories of the 1990s such as *Image and Logic* (Galison 1997) or *Lords of the Fly* (Kohler 1994). It's a question that hovers around the post-Kuhnian close inspection of tacit knowledge (Collins 2010), visual representation in science (Lynch and Woolgar 1990), laboratory notebooks (Holmes 1990), patents (Bowker 1992; Swanson 2007), and so on.

In my own work—particularly in conducting oral history interviews but also in reading through archival materials—I've tried keep that question front and center. And I pair with it a rule of thumb that's certainly fallible but perhaps still useful. Namely, if scientists and engineers spend a significant part of their day, or a significant part of their cognitive or emotional capacity, on X, then maybe X is important in the practice of science and engineering in their mutual shaping of (and by their society in the construction of) technical knowledge, and in the achievement of a variety of aims relative to scientists' and engineers' professional communities, home institutions, networks of personal affiliates, and segments of the societies they live in—even if, at the outset, X seems to have little to do with "Science" with a capital S.

So, for instance, if academic (and corporate and government) scientists and engineers spend a lot of time on teaching, well, maybe that's important. As David Kaiser and I (Mody and Kaiser 2008) have pointed out, most work in science studies ignores the pedagogical context in which much science takes place. That's perhaps in keeping with the ideology of science, as voiced, for instance, in Nobel lectures (Traweek 1988), where scientists often downplay their roles as teachers, students, and mentors. With explicit reference to Kuhn, though, Kaiser (2005a, b) has argued in his own work (and by editing and otherwise calling attention to the work of others) that teaching *is* important to science in many ways, not least in the creation of new knowledge. The classroom and the textbook and the mentoring relationship are sites where scientists advance arguments, where they discover and develop new ideas, and where they have a prime opportunity to engage their society's rather reasonable worries for the next generation's prospects.

Similarly, what if some scientists and engineers spend a lot of time writing popular books or consulting on blockbuster films? Well, maybe we should take that seriously as part of their efforts to achieve the aims of their normal science. Indeed, people like Gregg Mitman (1999) and David Kirby (2011) have done great work showing how, again, engagement with larger, popular audiences is not a distraction from scientific

---

[2] For an alternative reading of Kuhn's intentions, see Bird, Chap. 3, this volume

work. Rather, it *is* scientific work that offers researchers access to resources, a chance to try out ideas, to recruit new personnel, and to get a leg up in controversies with their colleagues—as well as a means to achieve important further aims, such as becoming famous.

What if we find that scientists and engineers spend a lot of time at conferences (Mody 2012; Ochs and Jacoby 1997)? I find this to be a weirdly neglected topic in science studies, since if the "social construction of knowledge" is observable, then it would have to be observable at scientific conferences—these, after all, are the occasions when scientists are both at their most social and most directly concerned with hashing out whose knowledge is correct. What if we find that scientists and engineers spend a lot of time traveling? That's a theme historians and anthropologists have picked up on for the field sciences (Kohler 2002; Helmreich 2009), but theoreticians and lab scientists travel as much as anyone. Why, and what do they accomplish by travel? What if we find that scientists and engineers spend a lot of time writing proposals or polishing appeals to their bosses or funders (Myers 1985)? Is that just a necessary evil, or is that what normal science *is* in a complex, heavily technological society where scientists and engineers make their normal science sustainable by attaching it to the concrete aims of politicians, bureaucrats, philanthropists, etc.? What if we find prominent scientists and engineers spending almost none of their time "in the lab," but instead managing their subordinates, traveling, writing grants, doing administrative work for the employers, etc. (Knorr-Cetina 1999)? Are they no longer "real" scientists or engineers? Or are they simply practicing their normal science by gathering resources and personnel and political goodwill?

More central to the present concerns of historians and sociologists of science is the discovery that many scientists spend much of their days participating in politics in one way or another—taking roles in government (Wang 1988), becoming activists (Moore 2008), pronouncing on the grand debates of their day (Egan 2007). Again, are such activities a distraction from "real" science? Or does Kuhn give us the tools to say that normal science is sustainable despite anomalies in its worldview partly because it achieves the aim of underwriting (or, occasionally, undermining) statecraft and political order? There are now several different keywords for describing that mutuality of science and politics. There's the "co-production" of knowledge and political order associated with Sheila Jasanoff (2004), sometimes with a nod to *Leviathan and the Air-Pump*. There's the Latourian actor-network (Latour 1987), in which the set of scientifically-known and technologically-made actors and actants is at the same time a kind of Parliament. There's the Foucauldian strand of science studies, probably best exemplified by Paul Rabinow (Rabinow and Dan-Cohen 2005), in which conceptions of self, morality, common sense are constantly remade by what we know and what we can build. And more! Science studies is now a field where the entanglement of social and scientific change is a given, in a way that Kuhn doesn't really gesture to in *Structure*, but which we would've been much slower to appreciate without him.

All that attention paid to the politics of/in science has to some extent distracted from other aspects of normal science, but there's still plenty to say. We now have some extraordinarily fine-grained studies of just how much normal science is shaped by completely ordinary activities that are largely conducted without any thought as to their scientific import. For instance, over the past twenty years or so, historians of science and technology have arrived at some rather counterintuitive conclusions from the mere fact that scientists and engineers eat and drink. I remember being warned in graduate school that the "what do scientists do all day" rule shouldn't be taken too far, since it couldn't possibly matter what a scientist ate for breakfast. Yet, as Steve Shapin and others (Lawrence and Shapin 1998) have shown, the scientist's credibility, and his or her ability to carve out a role of authority in their society, may indeed depend on what they eat or don't eat. One of the great discoveries of science studies is the degree to which scientific discovery and technological innovation may be the upshot of a conviviality that the participants may engage in for all kinds of other reasons – whether they be botanists in English pubs (Secord 1994), copier repair technicians in diners (Orr 1996), or electrical engineers creating Silicon Valley over ham radio sets or drinks at the Wagon Wheel bar (Lécuyer 2006). I would note with some pride that the index of my own history of scanning probe microsocpy (Mody 2011) lists two occurrences under the entry for "beer."

In a slightly different vein, Michael Lynch, Simon Cole, Ruth McNally, and Kathleen Jordan (Lynch et al. 2008) have coined the term "sub-normal" science to describe occupations that are manifestly technical—that require some sophisticated scientific expertise—and yet which are so routine as to be completely insulated from the possibility of revolution and paradigm shift. The particular case they have in mind is forensic science, especially DNA "fingerprinting," but the space of sub-normal science is quite large—possibly several times as large as the space of "normal" science. Think of all the national metrology institutes that set standards for every mundane food and substance (Lezaun 2012), or ministries of agriculture randomly testing crops and meat for disease or regulated substances (Lezaun 2006), or quality control laboratories in dairies and breweries. Sub-normal science deserves considerably more attention than it has received thus far, in part because, as Philip Mirowski (2011) has argued, it may be crowding out "normal" science. Sub-normal science is cheaper and more predictable than normal science, and therefore in some ways preferable to commercial concerns. Thus, firms in some industries—particularly in pharmaceuticals and other biomedical sectors—have outsourced more and more of their research to contract research organizations. If Mirowski is right, this shift poses a grave threat to normal science's capacity to unpredictably surprise and discover.

At the other end of the spectrum, we've also learned a great deal recently about all the different ways that seemingly abnormal science is in fact critical to the normal scientific enterprise. Historians and sociologists have found scientists and engineers spending much of their time and energy obsessed with seemingly quite unscientific— even purportedly antiscientific—ideas, and yet also enormously productive in the eyes of many of their contemporaries. For instance, Andrew Pickering (2010), Matt Wisnioski (2003), Thierry Bardini (2000), and others have shown how top-shelf researchers (psychologists, engineering scientists, mathematicians, etc.) immersed themselves in the countercultural world of drugs, mysticism, and avant-garde art in the late 1960s and 1970s. More recently, there have been a few notable cases in the field of nanotechnology of very prominent biologists and chemists espousing young-earth creationism, or something very close to it. Kaiser (2011), again, has recently shown that one of the most active and respected areas of physics research today—quantum entanglement and Bell's Theorem—was only rescued from the dustbin of disciplinary neglect by a rag-tag group of "hippie" physicists in the early 1970s because of their (to them) related interest in ESP, astral projection, time travel, UFOs, and communication with the dead.

Note that all this was happening in living memory, rather than in the time of Newton and Boyle! Ought we to 'tsk tsk', as previous generations did about Newton's alchemy and theology? Or should we take these activities and beliefs seriously as something some scientists and engineers care deeply about, that they see as integral to their conception of what science is and what purpose it serves, and that they mine for ideas, skills, resources, and connections?

Normal science, sub-normal science, abnormal-but-still-productive science—these and the other areas I've surveyed are all topics that have followed naturally from the research program set in motion by Kuhn—or, perhaps by those researchers who re-discovered Ludwik Fleck ([1935] 1979) as a result of Kuhn. However, there have been a few areas where Kuhn's preoccupations have hindered our appreciation of some pervasive and, frankly, quite important aspects of normal science. Let me raise two of these, one more in science studies and the other more in technology studies.

First, in science: if we apply the "what do scientists do all day" rule, we'll find, among other things, that a great many scientists and engineers spend a lot of their time taking out patents, starting companies, working for or consulting with firms, agitating for universities or government labs to found tech transfer offices, etc. There are many post-Kuhnians who would like to ignore those activities, as Kuhn did. There are a few loud voices in our field who decry such activities and who even find the study of such phenomena suspicious.

It seems to me, though, that the properly Kuhnian task, for our present age, is to ask what scientists and engineers (and the organizations that employ them) get out of such activities (especially since profit is often *not* one of the outcomes). As Steve Shapin (2008), Paul Rabinow (1996), and many others—including, I hope, myself in some way—have concluded, entrepreneurial science isn't *necessarily* problematic science. Indeed, engagement with the marketplace can enable certain grounds for creativity and persuasion that make for very good science. Industrial and entrepreneurial science weren't interesting to Kuhn, and that lack of interest is one of his less positive legacies that our field is still working through. But the study of industrial and entrepreneurial research, at least as I would encourage my colleagues to approach it, is very much in the vein of figuring out what *most* Kuhnian normal science is like.

Similarly, in technology studies, we're hampered by an obsession with technological revolution that isn't, of course, solely traceable to Kuhn but which Kuhn's work has fostered. One piece of evidence for this, I think, is the widespread use of "paradigm shift" as a buzzword in the business world. I'd point in particular to

David Edgerton (2007) as someone who's provocatively and *usually* correctly criticized science and technology studies for its infatuation with revolution. As Edgerton has argued, in science studies we talk too much about research at the expense of development, and in technology studies we talk too much about innovation at the expense of use and maintenance (and even when we do talk about use, we're usually focused on new or innovative users, rather than "normal" use).

Now, I'm not going to pretend that asking "what do scientists and engineers do all day" always yields interesting answers. There's as much time wastage in science as anything, so much of what scientists and engineers do all day has little bearing on anything. Though I suspect that time wastage in science might indeed play an important role in knowledge creation—there's a great study waiting to be done of boredom and idle time in science. I won't pretend, though, that we should *only* be asking what scientists and engineers do all day, since much of what any of us does is framed by a social order that rarely explicitly impinges on our actions and which we may not even be aware of. Nor do I pretend that the kinds of answers to that question that I've just outlined are very original—these are the kinds of things historians, sociologists, and anthropologists of science have been working through ever since Kuhn.

But if we keep that question at the ready, and if we push ourselves to revisit it in fresh ways, then we stand a chance of learning something about the structure of normal science. And that, I hope, puts science and technology studies in a better position to inform policies for science and technology, to better connect with all of the stakeholders in science and technology, to help scientists and engineers organize themselves more effectively on their own terms but also to be more engaged citizens. Ultimately, if we're still interested in pushing the Kuhnian project forward, then knowing more about what scientists and engineers do all day is fundamental to understanding the structure both of scientific revolutions and of scientific normalcy.

References

Ashmore, M., M. Mulkay, and T. Pinch 1989. *Health and efficiency: A sociology of health economics*. Milton Keynes: Open University Press.

Bardini, T. 2000. *Bootstrapping: Douglas Engelbart, coevolution, and the origins of personal computing*. Stanford: Stanford University Press.

Bowker, G. 1992. What's in a patent?. In *Shaping technology/building society: Studies in sociotechnical change*, eds. W. Bijker and J. Law, 53–75. Cambridge: MIT Press.

Collins, H. M. 1985. *Changing order: Replication and induction in scientific practice*. London: Sage.

Collins, H. M. 1990. *Artificial experts: Social knowledge and intelligent machines*. Cambridge: MIT Press.

Collins H. M. 2010. *Tacit and explicit knowledge*. Chicago: University of Chicago Press.

Collins, H. M., and T. J. Pinch 1982. *Frames of meaning: The social construction of extraordinary science*. London: Routledge.

Collins, H. M., and R. Evans 2002. The Third Wave of science studies: Studies of expertise and experience, *Social Studies of Science* 32:235–296.

Daston, L., and P. Galison 2007. *Objectivity*. New York: Zone Books.

Doing, P. 2009. *Velvet revolution at the synchrotron: Biology, physics, and change in science.* Cambridge: MIT Press.

Dreyfus, H. L. 1972. *What computers can't do: The limits of artificial intelligence.* New York: Harper and Row.

Dreyfus, H. L. 1992. *What computers still can't do: A critique of artificial reason.* Cambridge: MIT Press.

Edge, D. O., and M. J. Mulkay 1976. *Astronomy transformed: The emergence of radio astronomy in Britain.* New York: Wiley.

Edgerton, D. 2007. *The shock of the old: Technology and global history since 1900.* Oxford: Oxford University Press.

Egan, M. 2007. *Barry Commoner and the science of survival: The remaking of American environmentalism.* Cambridge: MIT Press.

Fleck, L. [1935] 1979. *Genesis and development of a scientific fact.* trans. F. Bradley and T. J. Trenn, ed. T. J. Trenn and R. K. Merton. Chicago: University of Chicago Press.

Galison, P. 1997. *Image and logic: A material culture of microphysics.* Chicago: University of Chicago Press.

Garfinkel, H., M. Lynch, and E. Livingston 1981. The work of a discovering science construed with materials from the optically discovered pulsar, *Philosophy of the Social Sciences* 11:131–158.

Gusterson, H. 1996. *Nuclear rites: A weapons laboratory at the end of the Cold War.* Berkeley: University of California Press.

Helmreich, S. 2009. *Alien ocean: Anthropological voyages in microbial seas.* Berkeley: University of California Press.

Holmes, F. L. 1990. Laboratory notebooks: Can the daily record illuminate the broader picture, *Proceedings of the American Philosophical Society* 134:349–366.

Jasanoff, S., ed. 2004. *States of knowledge: The co-production of science and social order.* London: Routledge.

Kaiser, D. 2005a. *Drawing theories apart: The dispersion of Feynman diagrams in postwar physics.* Chicago: University of Chicago Press.

Kaiser, D., ed. 2005b. *Pedagogy and the practice of science: Historical and contemporary perspectives.* Cambridge: MIT Press.

Kaiser, D. 2011. *How the hippies saved physics: Science, counterculture, and the quantum revival.* New York: W.W. Norton.

Kelty, C. 1997. The Whitehead Institute: A video portrait. Paper presented at the Society for Literature and Science meeting, Pittsburgh, PA, 30 October—2 November.

Kirby, D. 2011. *Lab coats in Hollywood: Science, scientists, and cinema.* Cambridge: MIT Press.

Knorr-Cetina, K. 1981. *The manufacture of knowledge: An essay on the constructivist and contextual nature of science.* Oxford: Pergamon Press.

Knorr-Cetina, K. 1999. *Epistemic cultures: How the sciences make knowledge.* Cambridge: Harvard University Press.

Koertge, N., ed. 1998. *A house built on sand: Exposing postmodernist myths about science.* New York: Oxford University Press.

Kohler, R. E. 1994. *Lords of the fly: Drosophila genetics and the experimental life.* Chicago: University of Chicago Press.

Kohler, R. E. 2002. *Landscapes and labscapes: Exploring the lab-field border in biology.* Chicago: University of Chicago Press.

Labinger, J., and H. M. Collins, eds. 2001. *The one culture? A conversation about science.* Chicago: University of Chicago Press.

Latour, B. 1987. *Science in action: How to follow scientists and engineers through society.* Cambridge: Harvard University Press.

Latour, B. 1999. Circulating reference: Sampling the soil in the Amazon rain forest. In *Pandora's hope: Essays on the reality of science studies,* 24–79. Cambridge: Harvard University Press.

Latour, B., and S. Woolgar 1986. *Laboratory life: The construction of scientific facts.* Princeton: Princeton University Press.

Lawrence, C., and S. Shapin eds. 1998. *Science incarnate: Historical embodiments of natural knowledge*. Chicago: University of Chicago Press.

Lécuyer, C. 2006. *Making Silicon Valley: Innovation and the growth of high tech, 1930–1970*. Cambridge: MIT Press.

Lezaun, J. 2006. Creating a new object of government: Making genetically modified organisms traceable, *Social Studies of Science* 36:499–531.

Lezaun, J. 2012. The pragmatic sanction of materials: Notes for an ethnography of legal substances, *Journal of Law and Society* 39:20–38.

Lynch, M. 1985. *Art and artifact in laboratory science: A study of shop work and shop talk in a research laboratory*. London: Routledge.

Lynch M., and S. Woolgar 1990. *Representation in scientific practice*. Cambridge: MIT Press.

Lynch, M., S. A. Cole, R. McNally, and K. Jordan 2008. *Truth machine: The contentious history of DNA fingerprinting*. Chicago: Chicago University Press.

Mackenzie, D. 1990. *Inventing accuracy: A historical sociology of nuclear missile guidance*. Cambridge: MIT Press.

Mirowski, P. 2011. *Science-mart: Privatizing American science*. Cambridge: Harvard University Press.

Mitman, G. 1999. *Reel nature: America's romance with wildlife on films*. Cambridge: Harvard University Press.

Mody, C. C. M. 2010. Fact and friction, *Metascience* 19:493–496.

Mody, C. C. M. 2011. *Instrumental community: Probe microscopy and the path to nanotechnology*. Cambridge: MIT Press.

Mody, C. C. M. 2012. Conferences and the emergence of nanoscience. In *The social life of nanotechnology,* ed. B.H. Harthorn and J. Mohr, 52–65. London: Routledge.

Mody, C. C. M., and D. Kaiser 2008. Scientific training and the creation of scientific knowledge. In *Handbook of science and technology studies,* ed. E. J. Hackett, O. Amsterdamska, M. Lynch, and J. Wajcman, 377–402. Cambridge: MIT Press.

Moore, K. 2008. *Disrupting science: social movements, American scientists, and the politics of the military, 1945–1975*. Princeton: Princeton University Press.

Myers, G. 1985. The social construction of two biologists' proposals, *Written Communication* 2:219–245.

Nickles, T. 2002. Normal science: From logic to case-based and model-based reasoning. In *Thomas Kuhn,* ed. T. Nickles, 142–177. Cambridge: Cambridge University Press.

Ochs, E., and S. Jacoby 1997. Down to the wire: The cultural clock of physicists and the discourse of consensus, *Language in Society* 26:479–505.

Orr, J. 1996. *Talking about machines: An ethnography of a modern job*. Ithaca: ILR Press.

Pickering, A., ed. 1992. *Science as practice and culture*. Chicago: University of Chicago Press.

Pickering, A. 1995. *The mangle of practice: Time, agency, and science*. Chicago: University of Chicago Press.

Pickering, A. 2010. *The cybernetic brain: Sketches of another future*. Chicago: University of Chicago Press.

Pinch, T. J. 1986. *Confronting nature: The sociology of solar-neutrino detection*. Dordrecht: Reidel.

Rabinow, P. 1996. *Making PCR: A story of biotechnology*. Chicago: University of Chicago Press.

Rabinow, P., and T. Dan-Cohen 2005. *Machine to make a future: Biotech chronicles*. Princeton: Princeton University Press.

Roland, A., and P. Shiman 2002. *Strategic computing: DARPA and the quest for machine intelligence, 1983–1993*. Cambridge: MIT Press.

Secord, A. 1994. Science in the pub: Artisan botanists in early nineteenth-century Lancashire, *History of Science* 32:269–315.

Shapin S. 1975. Phrenological knowledge and the social structure of early nineteenth-century Edinburgh, *Annals of Science* 32:219–243.

Shapin, S. 2008. *The scientific life: A moral history of a late modern vocation*. Chicago: University of Chicago Press.

Shapin S., and S. Schaffer 1985. *Leviathan and the air-pump: Hobbes, Boyle, and the experimental life*. Princeton: Princeton University Press.
Swanson, K. 2007. Biotech in court: A legal lesson in the unity of science, *Social Studies of Science* 37:357–384.
Traweek, S. 1988. *Beamtimes and lifetimes: The world of high energy physicists*. Cambridge: Harvard University Press.
Wang, Z. 2008. *In Sputnik's shadow: The President's Science Advisory Committee and Cold War America*. New Brunswick: Rutgers University Press.
Wisnioski, M. 2003. Inside 'the System': Engineers, scientists and the boundaries of social protest in the long 1960s, *History and Technology* 19:313–333.

Chapter 8
From Theory Choice to Theory Search: The Essential Tension Between Exploration and Exploitation in Science

Rogier De Langhe and Peter Rubbens

8.1 Introduction

Theory choice is one of the most important problems in philosophy of science. Some argue that the choice of one theory over another is rational if the procedure that led up to that choice was rational (e.g., Popper (1963) proposed a methodology based on falsifiability), while others have argued that choice for a theory is rational because of certain properties of that theory itself, called "scientific virtues" (e.g., Poincaré (1905)'s defence of the virtue of simplicity). However a standard assumption is that the theories from which to choose *already exist*. A typical assumption in twentieth century philosophy of science has been to restrict itself to the "context of justification", treating theories as given. (Reichenbach 1938) This assumption reduces the problem of theory choice to a problem of choice under *risk*. Although at the moment of choice it is unknown which theory will eventually be right, a comparison of theories' past performance allows to select the one which is most probable to be successful in the future. However, if no scientist ever chooses to search for a new theory, scientific progress would soon come to a halt. The starting point of this paper is therefore that at least some scientists must search for new theories rather than keep on developing the existing ones. By shifting the focus of scientific rationality from choice among given alternatives to finding a balance between the exploitation of existing theories and the exploration of new theories, the activity that scientists engage in is no longer one of passive choice but of active search. By framing the problem of theory choice as search, we argue that a more realistic thematization ensues of the problem that practicing scientists are confronted with: the question of whether to expand on existing theories or start working on an entirely new theory.

R. De Langhe (✉)
Tillburg Center for Logic and Philosophy of Science (TiLPS),
Tilburg University, Tilburg, The Netherlands
e-mail: rogierdelanghe@gmail.com

P. Rubbens
Department of Physics and Astronomy, Ghent University, Ghent, Belgium

© Springer International Publishing Switzerland 2015
W. J. Devlin, A. Bokulich (eds.), *Kuhn's Structure of Scientific Revolutions—50 Years On,*
Boston Studies in the Philosophy and History of Science 311,
DOI 10.1007/978-3-319-13383-6_8

This expansion of the problem of theory choice entails that the set of alternatives from which a choice must be made is no longer given but infinite. Despite the fact that the number of questions that can be asked about the world is infinite, we observe that scientists nevertheless work together without any centralized control and with only local information available. It is then all but a miracle that these independent scientists succeed in collective construction of theories. Conversely, once a collectively shared set of questions and methods has been accepted, it is difficult to see how individual agents could unilaterally succeed in changing it. To be successful, a community of scientists must therefore find the right balance between the exploitation of existing theories and the creation of new theories. The problem of theory choice then becomes one of theory search, and how scientists can rationally decide between both.

Expanding the problem of theory choice to theory search turns the problem of theory choice under risk into a problem of choice under *uncertainty*. Because the set of alternatives is infinite, the criteria a good theory must meet cannot be specified independently in advance. And because the properties of the new theories are unknown when the choice for their development is made, no algorithm can be specified for their selection. Uncertainty therefore entails analytical intractability. Nevertheless we know from many domains in life that successful action is still possible under uncertainty on the basis of heuristics. A heuristic only tells agents how to look, not what to find and thereby guides the decision process without determining it. It is less specified than an algorithm, but it is this lack of specificity which makes it robust against choice for unknown alternatives. As such, heuristics are not inferior to algorithms, but a different solution to a different problem.

Possibly it is this lack of algorithmic treatment which explains why philosophers have been so keen on maintaining the unrealistic restriction to the context of justification. Thomas Kuhn (1970) is a famous exception. In response to the ensuing uncertainty, Kuhn suggested that theory choice is based on heuristics rather than on an algorithm. He described five common scientific virtues (accuracy, consistency, scope, simplicity and fruitfulness) as "*criteria that influence decisions without specifying what those decisions must be.*" (Kuhn 1977, p. 330) However Kuhn was unable to specify how possibly such values could lead decentralized scientists to produce collectively successful science as we know it.[1]

Here we define a successful scientific community as a community which finds a rational balance between exploration and exploitation. Using the novel tool of agent based-modeling, we show that a succesful scientific community can emerge from agents choosing based on a simple heuristic applied to local information.

---

[1] "Even those who have followed me this far will want to know how a value-based enterprise of the sort I have described can develop as a science does, repeatedly producing powerful new techniques for prediction and control. To that question, unfortunately, I have no answer at all [. . . .] The lacuna is one I feel acutely" (Kuhn 1977, pp. 332–333).

Fig. 8.1 Moore
neighbourhood, $H = 8$.
Figure used from Polhill et al.
(2010)

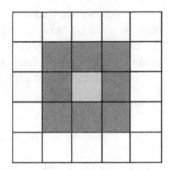

8.2   Heuristic: Exploitation and Exploration

Consider a community of $N(1, \ldots, n)$ scientists. Each turn, each scientist makes a contribution $C(c_1, \ldots, c_N)$ to a theory $S(s_1, \ldots, s_M)$; mark that N is a constant of the system, however, M may vary as the system evolves. A heuristic for individual theory search must rely only on information locally available to the agent and must be sufficiently general to be applicable across large ranges of possible choice situations. The balance between exploration and exploitation is a general consideration applicable to any conceivable alternative.

Exploitation consists in an allocation of scientific labor to an existing theory. The more scientists exploit the same theory, the higher the benefit of specialization becomes because scientists can specialize in narrower subproblems and specialized tools can be developed.[2] As a consequence, a local proxy for the benefits of exploitation is the number of adopters of a theory. More precisely, the "adoption" A of a theory s is the sum of the number of scientists that contribute to it, where H denotes the size of the local neighbourhood of a scientist.

$$A_s(t) = \sum_{i=0}^{H} a_{i,s}(t). \qquad (8.1)$$

In Fig. 8.1 an example of a local neighbourhood is given, a so-called *Moore neighbourhood*, for which $H = 8$.

Exploration consists in an allocation of scientific labor to a new theory. The less articulated a theory, the higher the innovative value of contributing to that theory. We will assume that each scientist makes one contribution at each turn. As a consequence, a local proxy for the benefits of exploration is the inverse of the number of contributions made to a theory. More precisely, the "production" P as the sum of contributions to a theory s is the sum of adopters through time t:

$$P_s(t) = \int_0^t A_s(t')dt'. \qquad (8.2)$$

---

[2] The insight that division of labor increases productivity by fostering specialization is as old as Adam Smith (1776, 2003) and marked the birth of modern economics.

The relation between adoption and production as specified here dynamically captures the trade-off between exploitation and exploration. More exploitation means less exploration, and similarly the number of adopters to a theory increases the specialization benefits from exploiting it but decreases the novelty of exploring it.

Since exploration and exploitation jointly determine the utility of a contribution to a theory, this utility can then be expressed as follows:

$$U_s(t) = \frac{A_s^\alpha(t)}{P_s(t)}. \tag{8.3}$$

The parameter α denotes the output elasticity of coordination, which is a function of the (for the purpose of this paper exogenous) *state of technology*. U_s is backward-looking because it evaluates the utility of the last contribution to a theory. But if scientists would always choose to develop that theory which is best developed, then soon enough no alternative theories would ever be developed because any new theory would always be less developed than the existing ones. Scientists can only be expected to develop new theories if their focus is not backward-looking, but forward-looking. Therefore the utility of the next contribution should be considered:

$$U_s'(t) = \frac{(A_s(t) + 1)^\alpha}{P_s(t) + 1}. \tag{8.4}$$

8.3   Dynamics: A Battle of Perspectives

In the face of unknown alternatives, scientists must take into account the actions of others. For an adopter of a theory, the trade-off between exploration and exploitation entails a tension in the value of a new adopter to that theory: in the short term it increases the number of agents with whom labor can be divided and therefore the exploitation part of the value of that theory, but in the long term it leads to a higher production within that theory and thereby decreases the exploratory value of contributions to that theory. However for each individual agent acting locally, the benefits of a new adopter outweighs their cost, and therefore agents try to persuade each other to adopter their theories.[3] As such we specify the dynamics of the model as driven by the desire of agents to persuade each other: every turn N randomly assigned agents are selected to be convinced by one of their neighbors. The probability that this convincing attempt is successful ("conversion") is proportional to the utility of the respective theories from the perspective the adopter of that theory.[4] As such there is what we call a "battle of perspectives" by which theories are confronted

---

[3] It is only in the long term that these individual decisions will collectively exhaust that theory and lead to its collapse.

[4] Each agent only knows the utility of his own theory based on his local information from the neighborhood and when a convincing attempt is made the probability of success is proportional to the utility of their respective theories.

based on the value the respective defender of that theory attributes to it. The battle of perspectives has three possible outcomes: stick to ones theory i (P_s), be convinced by the convincing neighbor's theory j (P_c) and contribute to an entirely new theory k (P_n). Our definition of the utility of a contribution to a theory (see Eq. 8.4) allows us to quantify the probability of each of these outcomes:[5]

$$P_s = \frac{U_{s_i}}{U_{s_i} + U_{s_j} + 1}, \tag{8.5}$$

$$P_c = \frac{U_{s_j}}{U_{s_i} + U_{s_j} + 1}, \tag{8.6}$$

$$P_n = \frac{1}{U_{s_i} + U_{s_j} + 1}. \tag{8.7}$$

8.4 Simulation Results

A heuristic for individual theory search was developed and the probabilities for the resulting dynamics were quantified. The only information lacking to understand the consequences of this model is that on adoption. Adoption in the model will vary as a consequence of the probabilities of the endogenously created theories, and those probabilities are in turn determined by adoption. As a consequence the system is permanantly out of equilibrium and its dynamics cannot be specified by detmerining equilibria but only by studying the resulting process. The study of aggregate patterns emerging from the interactions of individual agents is possible using the technique of agent-based modeling. Assume periodic boundary conditions and let agents interact on a two-dimensional grid with size L; this means we have $N = L^2$ agents. Each agent sees only his Moore-neighborhood and tracks each turn the number of adopters to each theory and their total production. Initially all agents adopt the same theory.

Figure 8.2 shows the evolution of adoption to the dominant theory in a typical run of the model. A few observations can already be made:

- Although there is only one initial theory, novel theories are created endogenously.
- Although the number of possible alternative theories is all but infinite, cooperation on a single theory emerges and novel theories are only created as existing ones are exhausted. The model thus self-organizes to find a dynamic balance between locking in to a single theory and a situation where there are as many theories as there are scientists.
- All action in the model is taken by individuals. The fact that theory change occurs proves that individuals have the capacity to unilaterally initiate theory change.

---

[5] Note that the utility of contributing to an entirely new theory is always exactly 1 because an undeveloped theory has no adopters and no production so both A and P are 0 and Eq. 8.4 always equals 1 irrespective of α.

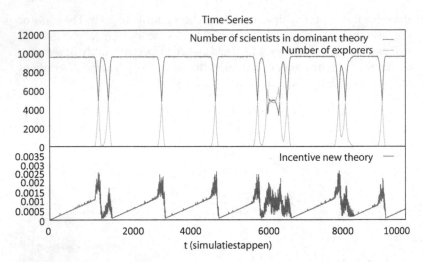

Fig. 8.2 Time-series in which the largest number of scientists who work within the same theory and the incentive of a new paradigm has been visualized for $L = 100$ and $\alpha = 7$

- Disruption of consensus ("crisis") is of varying size and length, although typically shorter than the length of consensus.

The system exhibits a cross-over from essentially competitive to cooperative as α increases. In Fig. 8.3 we show the average size of the dominant theory for various α and different dimensions of L. Variation of all sizes can be observed around $\alpha = 6.5$, perhaps suggesting critical point. Communities with α below this point are characterized by continuous presence of multiple competing theories, while communities in which α is higher are characterized by the alternating monopoly of a single theory separated by shorter periods of crisis.

The evolution of the model is driven by the ever-changing incentive structure for the three alternatives available to each agent in the model. In particular, the probability of contributing to a new theory is inversely proportional to the sum of the utility of contributing to the existing theories. As a consequence the incentive for agents to create new theories is not given but varies endogenously with the dynamics of the model. The bottom of Fig. 8.2 shows how the incentive structure of the model, represented by the utility of contributing to a new theory, coevolves with the very contributions it regulates: the longer the period of consensus, the higher the probability that agents create a new theory; conversely the probability of creating a new theory is lowest when consensus on a new theory has just emerged. The figure clearly shows two separate phases: a normal phase in which the incentive structure is stable and predictable, and a revolutionary phase in which the incentive structure is unstable and unpredictable.

It is clear that these two phases in the incentive structure result in very different ratios of explorers and exploiters. These respective phases are characterized by statistically different properties, allowing for a quantitative separation of these phases based on the distribution of their production. During a normal phase characterized

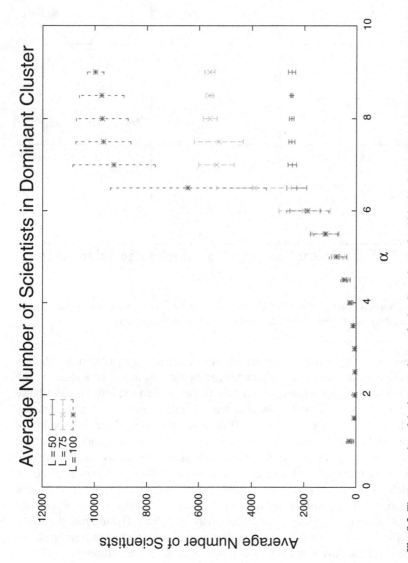

Fig. 8.3 The average size of the largest cluster of scientists who work within the theory that counts the most scientists for various α and different dimensions L

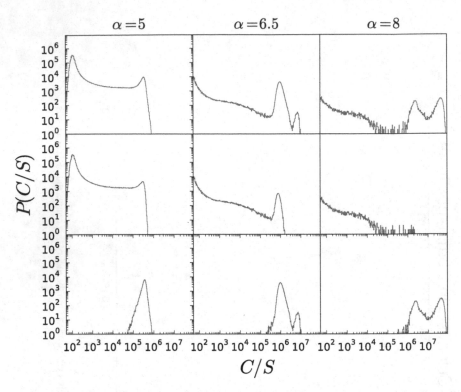

Fig. 8.4 Distribution of total production for $\alpha = 5, 6.5, 8$; $L = 100$. The first row visualizes all contributions, the second those of explorers and the third those of exploiters

by consensus on a single theory, production follows a markedly different distribution than during a revolutionary phase characterized by the absence of consensus.

Let an explorer be a contributor to a new theory and an exploiter a contributor to an existing theory. We will show that we are able to make a clear distinction between the two sorts of phases. We will do this by means of the distribution of the total production, defined as the total amount of contributions made to a certain theory s, visualized in Fig. 8.4. In this figure the distribution of the total number of contributions per theory S is visualized. This is done for three different values of α, one in which competitivity mainly determines the behavior of scientists ($\alpha = 5$), one in which coordination mainly determines the behavior of scientists ($\alpha = 8$) and one in which both determine the behavior of scientists ($\alpha = 6.5$). The top row shows the complete distribution of total production, the second row the production produced by explorers and the third row the production produced by the exploiters.

For $\alpha = 5$ we see that the contributions of exploiters have a marginal influence on the total distribution. However, they play a part in the dynamics, and it would be wrong to claim that there are no exploiters when α is low. When $\alpha = 6.5$, contributions of both explorers and exploiters matter to the total distribution. When looking at the productions of only explorers, it appears that they would sometimes also act in a coherent way. However, the bump can be explained by Fig. 8.3. For $\alpha = 6.5$, we see that we can have clusters of all sizes. This means that multiple large clusters are able to coexist, nevertheless, with our current definition of explorers, we do not take into account larger clusters who are not the largest cluster. These large clusters however give rise to the previously mentioned bump. Whether these scientists can be percieved as explorers or not is a discussion less important for the purposes of this paper.

When coordination is strong, which is the case for $\alpha = 8$, the distribution splits itself in two, in which the explorers are represented by the left distribution and the exploiters by the right. It is clear from this that explorers follow a different distribution than exploiters, which is globally visible only when the coordination is strong. From we this we can conclude that scientists are able to behave in two significantly different ways; either a scientist contributes to a well-established theory in which the gain is obvious because a lot of his peers are contributing to the same theory or a scientist contributes to a non-established en less-known theory, in which he explores the possibilities of that theory.

It is interesting to note that α plays an important role in visualizing both distributions. By gradually increasing α, the distribution of exploiters becomes more and more clear, and ultimately both distributions are visible and completely disconnected from each other.

Although scientists have the possibility to contribute to different theories, we can distinguish periods in which the whole scientific community unite themselves contributing to one theory. This happens in a self-organizing way, in which the model finds a dynamic balance between locking in to a single theory and a situation where there are as many theories as there are scientists. On top of that, new theories are created endogenously and only when the existing ones seem to get exhausted.

8.5 Conclusion

In conclusion, in the absence of centralized control, with only limited information and using nothing but a simple heuristic, the interactions of scientists result in a robust pattern of intermittent theory exploitation and exploration with shifts between them occurring at the rational point in time. Individual theory choice in this model is not a choice between existing alternatives, but a process of search which finds a dynamic balance between the actual and the possible; between given alternatives and creating new alternatives. As such the theories created define the possibility space for subsequent theory choice, all in the same model. It was shown in this paper that this heuristic results in a self-organizing balance between tradition and innovation, where theories are created as they are needed. From the interactions between individuals using only a local rule of thumb that strikes a balance between exploration and exploitation (both the benefits of specialization and the race for priority are taken into account) can emerge a community that finds a dynamic balance between the exploitation of existing theories and the exploration of new theories.

References

Kuhn, T. 1970. *The structure of scientific revolutions*. 2nd ed. Chicago: Chicago University Press.

Kuhn, T. 1977. *The essential tension*. Chicago: Chicago University Press.

Poincaré, H. 1905. *Science and hypothesis*. London: Walter Scott Publishing.

Polhill, J. G., Sutherland, L.-A., and Gotts, N. M. 2010. Using qualitative evidence to enhance an agent-based modelling system for studying land use change. *Journal of Artificial Societies and Social Simulation* 13 (2): 10. http://jasss.soc.surrey.ac.uk/13/2/10.html.

Popper, K. 1963. *Conjectures and refutations*. London: Routledge.

Reichenbach, H. 1938. *Experience and prediction*. University of Chicago Press.

Smith, A. 1776, 2003. *Wealth of nations*. New York: Bantam Classics.

Chapter 9
The Evolving Notion and Role of Kuhn's Incommensurability Thesis

James A. Marcum

9.1 Introduction

The incommensurability thesis (IT) played a vital role in Thomas Kuhn's philosophy of science. In 1962, Kuhn introduced IT in the *Structure of Scientific Revolutions* (*Structure*) to make sense of what he claimed are nonsensical statements in antiquated scientific texts. Briefly, the original incommensurability thesis (IT_O) states that after a scientific revolution, the old and new paradigms are incompatible or incommensurable to one another, i.e. no common measure or meaning exists by which to compare paradigms to each other. Thus, scientific communities subscribing to different paradigms are unable to communicate adequately, if at all, with one another.

The philosophy of science community was critical of IT_O, arguing it made science both irrational and relativistic (Wang 2007). In response, Kuhn proposed several modified versions of incommensurability, which reflected a "linguistic turn." The first version Kuhn (1970) introduced in a 1969 Postscript to the second edition of *Structure*. Postscript IT (IT_P) stressed the semantic nature of incommensurability in which he replaced gestalt switches with linguistic communities. The second version, which Kuhn (1983) formally called the local incommensurability thesis (IT_L), held that only a section or subset of the new paradigm is incommensurable with the old one. His intention was to differentiate IT_L from radical IT (IT_R), which claims global or extreme incommensurability exists between two competing or successive theories or paradigms (Hung 2006; Sankey 2000).

Shortly after the introduction of IT_L, Kuhn (1991; 2000) shifted from a historical philosophy of science (HPS) to an evolutionary philosophy of science (EPS), which represented an "evolutionary turn" (Marcum 2013). And, with the shift came a change in both the notion and role for incommensurability. He now defined incommensurability in terms of changes in the lexical taxonomy of a scientific specialty,

J. A. Marcum (✉)
Department of Philosophy, Baylor University, TX, 76798 Waco, USA
e-mail: James_Marcum@baylor.ed

© Springer International Publishing Switzerland 2015
W. J. Devlin, A. Bokulich (eds.), *Kuhn's Structure of Scientific Revolutions—50 Years On*,
Boston Studies in the Philosophy and History of Science 311,
DOI 10.1007/978-3-319-13383-6_9

and Howard Sankey (1998) denoted it as the taxonomic incommensurability thesis (IT_T). Moreover, incommensurability now functioned for Kuhn as a mechanism to isolate a community's lexicon from another's and as a means to underpin the notion of scientific progress as the proliferation of scientific specialties. In other words, as the taxonomical structure of the two lexicons become isolated from one another and thereby incommensurable, according to Kuhn, the new specialty's lexicon splits off from the parent specialty's lexicon. This process accounts for his notion of scientific progress as an increase in the number of scientific specialties after a revolution. Scientific progress, then, is akin to biological speciation, with IT_T serving as the isolation or selection mechanism.

In what follows, I reconstruct Kuhn's evolving notion and role for incommensurability and critically analyze how he employed IT_T in an EPS to address the ideas of scientific truth and reality.

9.2   Kuhn's Introduction of IT_O

Kuhn, along with Paul Feyerabend, is generally credited for introducing IT into the philosophy of science literature (Hoyningen-Huene 2005). In an interview with Skuli Sigurdsson, Kuhn recounted its origins in an attempt to understand a passage from Aristotle's *Physics*. "What Aristotle could be saying baffled me at first," acknowledged Kuhn, "until—and I remember the point vividly—I suddenly broke in and found a way to understand it, a way which made Aristotle's philosophy makes sense" (Sigurdsson 1990, p. 20). The reason Kuhn was baffled is that, as he explained in *Structure*, he lived in a modern world not just incompatible, but incommensurable, with Aristotle's world (Kuhn 1962, p. 102). IT_O functioned for Kuhn in terms of distinguishing between normal and revolutionary science, especially their respective progress. Whereas normal-science progress is cumulative, revolutionary-science progress is not. For the latter, progress often involves a substantial or radical break with the past in which proponents of two competing or successive paradigms are unable to communicate adequately—if at all—with each other.

For Kuhn (1962), not only do entities and concepts that constitute a paradigm change during a scientific revolution, but so does the nature of science itself. The new paradigmatic science, for example, may marginalize or even ignore particular problems and their solutions conducted under the egis of the old paradigm or even declare them "unscientific," according to Kuhn. Moreover, the resolution of anomalies is especially pertinent, since the old paradigm cannot resolve them while the new one can. Thus, two paradigmatic sciences are vastly different from one another and may even conflict with each other, especially disputes over anomalies.

In *Structure*, Kuhn discussed three reasons to account for IT_O, which explain why proponents of competing paradigms generally fail to communicate with one another across a revolutionary divide (Sankey and Hoyningen-Huene 2001). The first is that "proponents of competing paradigms will often disagree about the list of problems that any candidate for paradigm must resolve" (Kuhn 1962, p. 147). This reason goes to the root of Kuhn's philosophy of science and the notion of scientific progress. According to Kuhn, no paradigm completely or accurately explains the natural world. Eventually anomalies arise, which scientists laboring under a prevailing paradigm may initially ignore. If these anomalies persist, however, and if they become an impediment to the practice of normal science, then a crisis may ensue. During the crisis, competing paradigms are proposed, and the scientific community chooses the paradigm that not only resolves the anomalies but also offers greater promise for guiding future scientific practice. The list of problems, then, is substantially different between the old paradigm and its competitor. For the old paradigm, the list excludes or marginalizes anomalies, while for the competitor the list includes them. As a result, the nature of science itself changes after a paradigmatic shift.

The next reason emerges from the first in that—although successive paradigms share certain conceptual and experimental "elements" of scientific practice—the relationship of these elements is profoundly altered in the new paradigm as it resolves the anomalies, which ultimately contributes to a communication rift between the two paradigmatic communities. Proponents of the new paradigm—in order to resolve the anomalies facing the old paradigm—formulate alternative meanings and relationships of concepts from those of the old paradigm. Kuhn gives the example from the Copernican revolution in which "earth" meant no longer a fixed body, according to the Ptolemaic paradigm, but a moving one. In other words, it became a planet with profound implications for not only cosmology and the relationships of celestial bodies to each other, but also for anthropology and the relationship of humans to the cosmos (Kuhn 1957). Successful communication between members of competing paradigmatic communities requires understanding what each community means by its terms and the accompanying relationships among them.

As for the last reason, which represents a culmination of the first two reasons, not only are the list of problems and the scope of communication affected, but the world—which the two competing paradigmatic communities inhabit and in which they conduct science—is also changed. According to Kuhn, "proponents of competing paradigms practice their trades in different worlds" (1962, p. 149). He compared paradigm shifts to gestalt switches in which community members adopting the new paradigm see the world differently from those committed to the old paradigm. Instead of seeing a duck, for example, community members adopting an alternative paradigm now see a rabbit. However, Kuhn was quick to add that, during a scientific revolution or paradigm shift, scientists of competing paradigms cannot simply "see anything they please. Both are looking at the world," he insisted, "and what they look at has not changed. But in some areas they see different things," as Kuhn explained, "and they see them in different relations one to the other" (1962, p. 149). To use the duck-rabbit gestalt switch, a person looking at lines on a piece of paper sees a duck while another person looking at the same lines sees a rabbit.

Through a process similar to gestalt switches, Kuhn argued, members of two competing scientific communities see and commit to different paradigms. The members of the community occupying a new paradigmatic world solve anomalies impeding scientific practice for members of the community laboring in the old paradigmatic world. In fact, members of the old community may not only fail to solve the anomalies challenging it, but they may be unable or even refuse to see the solutions a new paradigmatic world provides them. Thus, the two worlds are incommensurable and communication between the two respective communities inhabiting them remains partial, unless members of old community enter into or "convert" to the new way of seeing the world.

In summary, Kuhn's IT_O was the result of a "hermeneutical turn" in which he struggled to make sense of or to give meaning to select passages in Aristotle's *Physics* in terms of Newtonian mechanics. What he discovered as he labored to understand these passages is that he had to embrace Aristotle's worldview and its assumptions—and not impose those of Newton on the passages—before he could make sense of them. Incommensurability, for Kuhn in *Structure*, then, represented "no common meaning" between two competing or successive paradigms. As such, IT_O served a function in accounting for scientific progress during paradigm shifts or scientific revolutions. In contrast to scientific progress that is cumulative during the practice of normal science, scientific revolutions involve a break with a previous paradigm such that the two paradigms are incommensurable.

9.3   Kuhn's Shift from IT_O to IT_P

Criticism of Kuhn's IT_O came predominantly from the philosophy of science community. That community leveled two major criticisms against IT_O. The first was the charge of irrationalism, i.e. because incommensurability precludes evaluative criteria independent of a given paradigm, critics saw Kuhn's notion of theory choice as an irrational process (Siegel 1987). Israel Scheffler, in an influential 1967 critique of *Structure* championed this criticism, especially given the fact that incommensurability impedes meaningful debate between scientific communities committed to competing paradigms and forces scientists to choose, not because of good reasons but because of persuasion. "Paradigm debates cannot, then," Scheffler asserted, "be understood in terms of categories of rational argument. They must fail to make logical or cognitive sense," he insisted, "owing to a fundamental failure of translation, and hence, of communication" (1982, p. 82). Since two competing paradigms are incommensurable with no recourse to paradigm-independent criteria for evaluating them, paradigm choice cannot be rational or based on good reasons. Rather, paradigm choice is dependent on non-rational faculties or no reason at all, and it is thereby an irrational or unreasonable process.

The second major criticism was relativism. In a 1964 review of *Structure*, Dudley Shapere contended that the ambiguous nature of Kuhn's notion of paradigm leads to a relativistic position. As Shapere reasoned:

> If one holds, without careful qualification, that the world is seen and interpreted 'through' a paradigm, or that theories are 'incommensurable,' or that there is 'meaning variance' between theories, or that all statements of facts are 'theory-laden,' then one may be led all too readily into relativism with regard to the development of science... such relativism... is

a logical outgrowth of conceptual confusions, in Kuhn's case. . . owing primarily to the use of a blanket term. (Shapere 1964, p. 393)

In other words, a scientific revolution is dependent on, or relative to, the particular paradigm to which a community subscribes, since paradigms define the nature of science itself and successive paradigms cannot be compared to each other directly.

The charges of irrationality and relativism against Kuhn's IT_O were also made at an international colloquium held in London in 1965. As for the irrationality charge, Imre Lakatos claimed that Kuhn's notion of scientific revolution and its associated IT_O have more to do with psychology than with epistemology. Since no rational standards exist external to a paradigm for evaluating competing paradigms and since the assent of the scientific community is the final arbitrator in paradigm selection, then, according to Lakatos, "in Kuhn's view scientific revolution is irrational, a matter of mob psychology" (1970, p. 178). As for the relativism charge, Karl Popper argued that Kuhn's IT_O suffered from "historical relativism" in which scientists espousing competing theories are "prisoners" trapped within a particular "framework" from which they cannot escape and consequently cannot compare their theories. Although Popper held a similar view, it differed from Kuhn's "in a Pickwickian sense: if we try, we can break out of our framework at any time" (1970, p. 56). In other words, Popper's position was that science is a critical activity in which competing theories can be compared with each other in a non-relativist manner, in contrast to Kuhn's relativism in which scientists are entrapped paradigmatically and cannot compare competing paradigms rationally because of incommensurability so that scientific progress is relative to change simply in paradigms.

Kuhn took seriously the charges of irrationalism and relativism against IT_O. He responded to these criticisms in a 1969 Postscript with IT_P. Although he admitted the ambiguity of paradigm as he used it in *Structure*, Kuhn claimed his critics, however, "misconstructed" IT_O as irrational. To defend incommensurability from the charge, he clarified the notion of paradigm by distinguishing between disciplinary matrices and exemplars. Disciplinary matrix refers to the milieu of a scientific community's practice, such as models, symbolic generalizations, and values, while exemplars represent "concrete problem-solutions" (1970, p. 187).

With a clarified notion of paradigm, Kuhn first defended incommensurability in terms of a scientific community's adherence to particular values. He argued that crisis resolution is not irrational simply because a community cannot formulate it using a "neutral algorithm." Rather, crisis resolution depends on a community's shared set of values, such as accuracy, simplicity, and fruitfulness. For example, a community might emphasize the value of fruitfulness to choose between two competing paradigms, while another simplicity. For Kuhn, incommensurability does not make paradigm choice irrational because no good reasons come into play; rather, it points to the fact that competing communities do not equally share a given set of values when evaluating competing paradigms. In other words, incommensurability is defined no longer as "no common meaning" but rather as "no common values."

Besides the values associated with a scientific community's disciplinary matrix, Kuhn also employed the notion of "similarity set" to defend incommensurability from the charge of irrationalism. He gave an example of the change in the grouping of planetary bodies after the Copernican revolution. For Kuhn, a scientific community participating in a crisis might utilize a new "similarity set" to resolve the crisis that differs significantly from another community's, especially if the new set resolves the anomaly or anomalies instigating the crisis. According to Kuhn, when changes in similarity sets take place during crisis resolution, "two men whose discourse had previously proceeded with apparently full understanding may suddenly find themselves responding to the same stimulus with incompatible descriptions and generalizations" (1970, p. 201). For Kuhn, changing similarity sets is not an irrational process but depends on a community's negotiations in composing these sets. IT_P, then, can be articulated not only as "no common values" but also as "no common exemplars."

In the Postscript, Kuhn also underwent a "linguistic turn" in defending incommensurability against the charge of irrationalism (Gattei 2008). He began by acknowledging that scientists as members of linguistic communities engaged in a crisis or "communication breakdown" do share certain common features, such as sensory stimuli and neural networks and programming. Given these features, Kuhn posited that members of each community must become translators. Beginning with terms that they share, each member of the communities representing a competing paradigm then identifies terms not shared with other communities. "Having isolated such areas of difficulty in scientific communication," as Kuhn expounded the ensuing step, "they can next resort to their shared everyday vocabularies in an effort to further elucidate their troubles" (1970, p. 202). The result is an ability of a member within a particular paradigm community to translate the alternative paradigm into that member's language. For Kuhn, then, translation provides entry into the world of an alternative paradigm in which a scientist is first persuaded to consider seriously and then converted by inhabiting it—or by going "native" as Kuhn articulated the process. Incommensurability, consequently, is not the irrational process as critics charged; rather, it is the essential feature by which science progresses reasonably.

Finally, having defended incommensurability from the charge of irrationalism, Kuhn proceeded—based on that defense—to support it against the charge of relativism (Bird 2011). He built a defense on the predominant criterion or value of any scientific community, "puzzle-solving ability" (1970, p. 205). Although application of puzzle solving might be "equivocal," Kuhn insisted "the behavior of a community which makes it preeminent will be very different from that of one which does not" (1970, p. 205). The difference between the two communities is that the one valuing puzzle-solving activity is not prone to what Kuhn considered "mere" relativism, while the other is. He then compared the development of a community's puzzle-solving activity to the evolution of biological species. Based on this comparison, Kuhn claimed that "scientific development is, like biological, a unidirectional and irreversible process. Later scientific theories are better than earlier ones for solving puzzles in the often quite different environments to which they are applied. That is not" he concluded, "a relativist's position" (1970, p. 206). But, if critics still want to make the charge that incommensurability functions relativistically because scientists cannot discover the truth about reality, Kuhn deemed the charge inconsequential

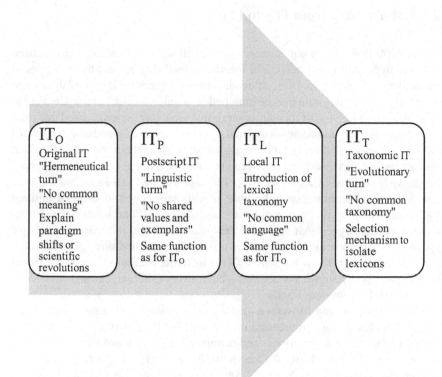

Fig. 9.1 Evolution of Kuhn's notion and role for IT

since he "cannot see that the relativist loses anything needed to account for the nature and development of the science" (1970, p. 207).

In summary, the shift in Kuhn's notion of commensurability from IT_O to IT_P was the result of a "linguistic turn" in which he abandoned the visual metaphor of gestalt switch for one of linguistic communities (see Fig. 9.1).

Interestingly, Peter Barker conjectures that Kuhn's use of gestalt switch was "too successful" in explaining incommensurability and was responsible for critics misreading Kuhn as advocating IT_R (2001, p. 437). Moreover, Kuhn also abandoned the notion of paradigm for the concepts of disciplinary matrix and exemplar and then defined incommensurability in terms of "no common values and exemplars" instead of "no common meaning." As for the role of incommensurability, Kuhn continued to use IT_P to account for scientific revolutions.

9.4   Kuhn's Shift from IT_P to IT_L

After the 1969 Postscript's publication, critics challenged both Kuhn's clarification of the paradigm notion and defense of incommensurability against their charges of irrationalism and relativism. In an influential review, Shapere still leveled the charge of paradigm ambiguity, although he admitted now only residual ambiguity, in his critique of Kuhn's defense of incommensurability. However, he maintained that even with just the residual ambiguity associated with the notion of paradigm in terms of deciding the meaning of terms, "it is impossible to be clear about the extent to which meanings determined by one paradigm can be expressed in the language of another" (Shapere 1971, p. 708). Alan Musgrave (1971) also asserted that Kuhn was not only unable to defend incommensurability, but he also eviscerated its original militant sting. Acknowledging that Kuhn did not subscribe to IT_R and its associated relativism, Musgrave argued that Kuhn has drastically demoted incommensurability's scope such that communication breakdown is easily remedied through translation. He concluded that, in Kuhn's defense of IT_P, incommensurability became a trivial notion. In sum, both reviewers concluded—but for very different reasons—that Kuhn failed to defend incommensurability.

Although Shapere and Musgrave, and others, certainly raised important issues with Kuhn's defense of incommensurability in the 1969 Postscript, Donald Davidson provided probably the most critical and certainly the best known and widely cited critique. In the 1973 Presidential Address to the American Philosophical Association, Davidson challenged the existence of "conceptual schemes" (aka Kuhn's paradigm concept) and its associated notion of incommensurability. He began with several descriptions of conceptual schemes, such as "ways of organizing experience" or "systems of categories that give form to the data of sensation" (1974, p. 5). He then pointed out that these schemes may reflect different realities such that what is real in one scheme may not be another. According to Davidson, these different schemes result in conceptual relativism, which leads to the following paradox: "Different points of view make sense, but only if there is a common coordinate system on which to plot them; yet the existence of a common system belies the claim of dramatic incomparability" (1974, p. 6). In other words, to claim two conceptual schemes are comparable or translatable and yet incommensurable is a contradiction.

In a paper delivered at the 1982 meeting of the Philosophy of Science Association, Kuhn again defended incommensurability from critics. To that end, he identified two lines of criticism concerning it and for both he cited Davidson's presidential lecture. The first assumes that some common ground must exist by which to compare competing theories; however, proponents of incommensurability deny this assumption and yet compare competing theories. Kuhn concluded that critics are charging proponents with being inconsistent in their use of incommensurability. The second line of criticism, as Kuhn articulated it, states,

> People like Kuhn, it is said, tell us that it is impossible to translate old theories into a modern language. But they then proceed to do exactly that, reconstructing Aristotle's or Newton's or Lavoisier's or Maxwell's theory without departing from the language they and

we speak every day. What can they mean, under these circumstances, when they speak about incommensurability? (Kuhn 1983, p. 670)

The charge, according to Kuhn, is that proponents of incommensurability are incoherent in their meaning of the concept. What is of interest to note is the shift in Kuhn's defense of incommensurability. No longer is he defending it against the charge that the proponents of incommensurability portray science as irrational or relativistic, but rather now he is defending these proponents from being inconsistent or incoherent, and he is offering an answer to what they mean by incommensurability.

Although Kuhn appreciated the seriousness of the charge raised by the first line of criticism, his main objective was to defend IT_P from the second line, i.e. to offer an intelligible and defensible notion of incommensurability. In other words, he answered critics like Davidson by defining what incommensurability is or means. To that end, he asserted that Davidson's criticism of incommensurability suffers from a fatal flaw—conflating interpretation with translation. In other words, Davidson relied on a false premise that translation is the same as interpretation—a premise Kuhn rejected.

According to Kuhn, translation involves a person who knows two languages and can thereby communicate intelligibly what is stated within a text in the language of another. In other words, terms exist in both languages to capture plainly what is expressed in either language. Interpretation, however, may involve a person who knows only one language but strives to make intelligible terms, like Willard Quine's "gavagai," that are initially unintelligible. To make such terms intelligible, an interpreter learns their meaning just as other terms in the interpreter's language were learnt originally. Importantly for Kuhn, if an unintelligible term like "gavagai" cannot be expressed in terms of an interpreter's language, i.e. it is "an irreducibly native term," then the interpreter must learn it with respect to the "structure" of its native world. "Those are the circumstances," concluded Kuhn, "for which I would reserve the term 'incommensurability'" (1983, p. 673).

To clarify the notion of incommensurability and to defend what he considered his initial intention for it, Kuhn (1983) introduced IT_L. He noted that, since many terms are common between two competing theories, one theory can be translated into another, and comparison between them is thereby possible. However, according to Kuhn, "for a small subgroup of (usually interdefined) terms and for sentences containing them do problems of translatability arise" (1983, pp. 670–671). Given IT_L, a historian like Kuhn in making sense of incommensurable theories, must interpret such terms by learning what the old term means *vis-à-vis* the "lexical network" in which it is embedded. He gave the example in which specific terms of an old theory like Newton's notion of mass cannot be translated with respect to a new one like Einstein's notion of mass (even though the term itself remains the same). For Kuhn, resolution of anomalies means changes in the network. He admitted IT_L is more "modest" than critics have ascribed to him, but the version represents his original intention. And, Kuhn insisted that even though IT_L represents a chastened version of incommensurability, still "communication ceases until one party acquires the language of the other" (1983, p. 683).

In summary, the shift in Kuhn's notion of commensurability from IT_P to IT_L was the result of further development in the "linguistic turn" he initiated in the 1969 Postscript, especially with the introduction of the notion of a scientific community's lexicon that replaced the paradigm concept and its predecessors, disciplinary matrix and exemplars (see Fig. 9.1). He explained incommensurability in terms of taxonomic categories, which do not overlap across a revolutionary divide. With this move, Kuhn now defined incommensurability in terms of "no common language" instead of "no common values and exemplars." As for the role of incommensurability, however, IT_L continued to function similarly to the role for IT_O and IT_P in accounting for scientific revolutions.

9.5   Kuhn's Shift from IT_L to IT_T

As the 1980s progressed, Kuhn strove to clarify the notion and role of incommensurability, with respect to the recent shift from a HPS (historical philosophy of science) to an EPS (evolutionary philosophy of science). To that end, he continued, in a series of lectures, to develop incommensurability in terms of a community's lexicon. In the 1984 Thalheimer lectures delivered at the Johns Hopkins, Kuhn discussed the nature of a scientific community's lexicon and the process by which its members learn it. What community members learn through acquiring a lexicon is a taxonomy that constitutes or stipulates the world's composition. And, the process of lexicon acquisition is holistic in that the taxonomy and its referents are learned together as a set. Consequently, incommensurability depends on lexical structure and its taxonomic categories. "Elements in the lexicon," Kuhn told the audience, "are thus linked together in such a way that changes in one cannot be made without changing others as well" (1984, p. 59). In other words, incommensurability reflects changes in particular taxonomic categories of the lexicon—especially changes precipitated by anomalies. Kuhn then introduced the notion of "feature space" in which terms are grouped together within a particular region of the lexicon and the notion of "similarity/difference metric," which represents the distance between these spaces. Incommensurability reflects the relative distance among feature spaces, so that the greater the "difference" metric, for instance, the more difficult communication is between members of incommensurable paradigms. Moreover, although Kuhn acknowledged that he is defending a "weaker form" of incommensurability, still, two incommensurable lexicons are "not fully compatible" and complete communication requires bilingualism.

Kuhn continued in a series of subsequent lectures to develop incommensurability towards its final version of IT_T. In a lecture delivered at the 1986 Nobel Symposium, Kuhn identified the following theme of his historical analysis of scientific revolutions: as scientific research continues, "a lexicon which gives access to one set of possible worlds also bars access to others" (1989a, p. 24). Again, he reiterated this theme in an ensuing lecture delivered at a conference sponsored by the Minnesota Center for Philosophy of Science. The lexicon, as "a structured vocabulary," provides not only entrance to a particular world but also the necessary guidance to investigate it. Importantly, for Kuhn, the lexicon is composed of "taxonomic categories" that represent a "taxonomic system" (1990a, pp. 314–315). Finally, in the 1987 Sherman Lectures delivered at the University College of London, he began to move away from the vagueness of the "language change" notion to the precision of "conceptual vocabulary" in a "structured conceptual lexicon" to defend his notion of world changes (Kuhn 1987). In these lectures, Kuhn was also beginning to identify an alternative role for incommensurability with respect to segregating or isolating lexicons and their associated worlds from one another. However, he still articulated incommensurability in terms of "no common language," with its attendant problems involving the notion of meaning, and had not fully transformed it with respect to an EPS.

Besides clarifying the notion and role of incommensurability during the 1980s, Kuhn also aggressively pursued the shift from HPS to EPS (Marcum 2013; Wray 2011). With the shift emerged a final version of incommensurability, IT_T, and a new role for it in the development of scientific knowledge and progress. Although Kuhn initially acknowledged in *Structure* the benefit of an evolutionary epistemology for articulating the development of science, he did not earnestly appropriate it until the late 1980s to early 1990s. Briefly, he proposed that scientific progress is comparable to biological evolution, with the emergence of new scientific specialties akin to speciation. The result is a tree-like structure with increased specialization at the tips of the branches. Lastly, Kuhn's EPS was non-teleological in the sense that science progresses not towards an ultimate truth about the world but simply away from a paradigm or theory that could not solve its anomalies to one that can.

Kuhn was working on a sequel to *Structure* to address several philosophical issues, especially incommensurability, which he raised in *Structure* without resolving. In a grant application to the National Science Foundation (NSF), he identified the sequel's "working title" as, *Words and Worlds: An Evolutionary View of Scientific Development* (Kuhn 1989b). It consisted of three parts, with three chapters in each part. In the first part, he explored the difficulties accessing or "breaking into" past scientific achievements. He concluded the first part, tethering incommensurability to a scientific community's lexicon. According to Kuhn, the problem or difficulty accessing past scientific achievements is that "alteration of word-meanings" represents "an alteration of the taxonomy embodied in the referring terms of a language" (1989b, p. 5). Thus, the lexical terms referring to objects change as the number of scientific specialties proliferate. In the next part, he discussed the lexicon's cognitive content with respect to taxonomic categories, composed of "contrast sets" in which entities do not overlap with one another—what he called the "no overlap principle," which prohibits the reference of terms to objects unless they are related to one another taxonomically. In the final part, he concluded the book by examining how lexicons change and the implications of this change for the ideas of scientific progress and realism.[1]

---

[1] For a further discussion of Kuhn's unfinished sequel to *Structure*, see Hoyningen-Huene, Chap. 13, this volume.

In the NSF grant application, Kuhn (1989b) outlined an evolutionary framework for scientific progress but he struggled to identify a role for incommensurability. To that end, he reframed incommensurability as a translatability problem instead of a meaning-variance problem. Invoking the notion of taxonomy, he argued that communication within a community of specialists depends on a shared taxonomic structure, which binds its members together. Kuhn next introduced the idea that a selection mechanism must operate in forming the community, just as natural selection operates in speciation during biological evolution. However, he was unable to identify what the mechanism is, although he felt that puzzle-solving, especially esoteric or anomalous puzzle-solving at the periphery of normal science practice, might provide a clue to identifying the selection mechanism. As Kuhn concluded, "if talk of 'puzzle solving' catches something about the selective mechanism which directs scientific advance, then it may provide a way to think usefully about the circumstances likely to foster or to inhibit science's further development" (1989b, p. 10). He was on the brink of solving his own puzzle about the role of incommensurability in the evolution of scientific knowledge.

Kuhn proposed a solution to the puzzle of a role for incommensurability in scientific progress in a pair of lectures delivered at UCLA in April 1990. To that end, he expanded the notion of taxonomic structure to include hierarchies besides simply categories. Thus, the relationship of terms within a lexicon is not simply one-dimensional, with respect to interactions such as similarity-difference relations among terms at a particular level, but it is also two-dimensional, with respect to inter-actions among different hierarchical levels. He illustrated a taxonomic hierarchy with a diagram of waterfowl. The higher level or "node," "animals," within the hierarchy exhibit certain features such as "feather," beaks," and "number of legs." Members of a particular language or scientific community utilize this "feature space" to classify physical objects. Kuhn also introduced the notion of "salience indices" to afford a fuller account of a taxonomy hierarchy. These indices, according to Kuhn, "provide the coordinates of a sort of center-of-gravity for the cluster of objects falling under that node within the space of differential features associated with the node above" (1990b, p. 6). For example, the features of "web feet" and "beak size" are salient for identifying birds *vis-à-vis* the features of the animal node. Kuhn specified, with the notions of "feature space" and "salience indices," a lexicon's taxonomic struc-ture. "People who share structure, also share meanings," concluded Kuhn; but, "if structure is not shared," he continued, "then translation breaks down" (1990b, p. 7). Thus, incommensurability reflects taxonomic systems of competing lexicons that classify referents and their referring terms differently with respect to feature space and salience indices and thereby results in segments of the respective lexicons that are untranslatable.

Kuhn then specified a new role for IT_T. In *Structure*, as he reminded the audi-ence, incommensurability functioned to account for the distinction between progress in normal science (accumulation to current paradigm) and in revolutionary science (rejection of an old paradigm and acceptance of a new one). In other words, incom-mensurability's role for a historian of science was to make sense of the seemingly incomprehensible antiquated scientific texts, after a paradigm shift or scientific revolution. In the sequel to *Structure*, Kuhn informed the audience, "it's a distinction between developments that can occur without revision of taxonomy and those that require local taxonomic changes" (1990b, p. 7). Thus, incommensurability's role for the historian of science is to identify specific taxonomic alterations of lexical structure in terms of feature space and salience indices, which account for the proliferation of scientific specialties after a revolution. To that end, Kuhn assigned incommensurability the function of isolating the lexicons of evolving scientific specialties to permit full development or speciation of the new specialty. Instead of its characteristic negative role of prohibiting communication between two incommensurable specialties or confusing modern historians when reading antiquated scientific texts, he now ascribed a positive role to it. "The breakdown of communication provides, I think," Kuhn proposed, "the isolating mechanism which promotes speciation, specialization, and which thus permits science to solve new puzzles with such effectiveness" (1990b, p. 9). Incommensurability plays a critical role in scientific progress by providing an opportunity for a new specialty's lexicon to develop fully without interference from the lexicon of the parent specialty.

In summary, the shift in Kuhn's notion of commensurability from IT_L to IT_T involved an "evolutionary turn" in which he converted from a HPS to an EPS (see Fig. 9.1). In addition, he advanced the language metaphor or the semantic dimension of incommensurability. The result was an explication of incommensurability in terms of "no common taxonomy" instead of "no common language." He also developed the concept of taxonomic system or network to include, besides categories, hierarchies. Finally, incommensurability in Kuhn's IT_T functions as an isolation mechanism that permits the lexicon of a new scientific specialty to develop as it splits off from the parent specialty.

9.6   Truth and Reality

With a robust version of EPS in hand that emphasized incommensurability's central role in scientific progress, Kuhn returned at the end of the UCLA lectures to address the issues of truth and reality. With respect to truth, he rejected the correspondence theory, i.e. scientific progress results in a better or truer understanding or mapping of reality. For Kuhn, the better scientific theory is not getting closer to the truth but "producing better and better instruments for solving problems and puzzles at the interface between man and nature" (1990b, p. 6). In other words, scientific progress results not in true or even truer statements about the world but in technology and instrumentation that afford evidence to solve problems relevant to a community of practitioners. Truth, according to Kuhn, functions logically as the rule of non-contradiction, "to force a choice between acceptance and rejection of a statement or a theory in the face of evidence shared by all" (1990b, p. 9). The idea of truth involves not the veracity or falsity of a statement; rather, it is involved in the evaluative process for accepting or rejecting a theory. In other words, truth plays simply an instrumental role in theory choice, i.e. theories are chosen not because their statements are true but because they do not contradict one another.

For Kuhn, truth is not the point of scientific practice and its progress. The point is that the lexicon determines such practice and progress. "And lexicons," claimed Kuhn, "are not the sorts of things that can be candidates for true/false. Rather, they're prerequisites for the statements and beliefs that are candidates" (1990b, p. 11). In other words, the lexicon specifies the possibility or conditions for conducting science and for guiding a community's puzzle-solving activity as it investigates the world. If truth does not drive scientific practice and progress, then what does—Kuhn asked the audience rhetorically. His answer was incommensurability. For incommensurability establishes communication within a scientific specialty by isolating the community and thereby allowing it to forge a lexicon that represents the world it encounters as it advances in its practice. For Kuhn, what is essential to progress in science is open communication within a scientific specialty, as it solves the puzzles—especially anomalies—facing it.[2]

Finally, Kuhn addressed the nature of reality and realism. He began his comments reminding the audience of the relationship between a community's lexicon and the world it inhabits, and on the importance that communication remains open for the opportunity of a community to progress in its practice. "Cognitive evolution," as Kuhn succinctly stated it, "depends upon exchange of statements within a community" (1990b, p. 13). He then compared biological evolution and the adaptation of organisms to a niche, to cognitive evolution. Just as biological organisms adapt to a niche, so the practice of a scientific specialty under the aegis of its lexicon leads to "closer and closer adaptation to a narrower and narrower niche" (1990b, p. 13). The lexicon itself is also a result of the adaptive process. Thus, a close link exists between the lexicon's "word" and the nature of the "world" it endeavors to describe—as evident from the working title of the projected sequel to *Structure*, *Words and worlds*. However, the world to which scientists and their lexical words adapt is not immaterial or simply a mental construct. According to Kuhn, "the world with which any community interacts through or by means of its lexicon is solid" (1990b, p. 13). He went on to give a trivial example of stubbing one's toe. "Community members can't simply decide how they'd like the world to be," concluded Kuhn, "and then enforce it" (1990b, p. 14). Consequently, he rejected the social constructionist's position that scientists manufacture or produce the world through their scientific practice and technical manipulations.

As Kuhn informed the audience, the position he advocates is not very different from the traditional idea of realism. "Knowledge of nature," asserted Kuhn, "is as firmly grounded as ever in rational deliberation about the results of experience" (1990b, p. 14). But, he promptly acknowledged that his position does differ from traditional realism on one count. "The world I've been speaking of is," claimed Kuhn, "lexicon dependent" (1990b, p. 14). Hence, the solidity of the world relies on the lexicon in that "given appropriate observational efforts, the world made available by

---

[2] For a further discussion of Kuhn and truth, see Devlin, Chap. 11, in this volume.

the lexicon will force consensus about the truth, assertability, facticity of statements about that world" (1990b, p. 14). He admitted that this position is relative in certain respects, especially in terms of lexical statements. Kuhn emphasized that what is relative, however, is not truth with respect to the "true/false game," as he called it, but the "effability, expressability, sayability" of these statements (1990b, p. 14). "And," he quickly added, "about what can't be said, questions of truth and falsity don't arise" (1990b, p. 14). Instead, lexicon-dependent statements exhibit an aesthetic quality that is truth independent.

Kuhn then broached candidly the topic of whether a mind-independent world exists, i.e. something not depending on a lexicon. Kuhn admitted something does exist "that provides the world's solidity, and grounds the true/false game. But so far as I can see," he admitted, "about that something, there's nothing descriptive that can be said" (1990b, p. 14). Hence, reality is not directly knowable or even utterable without reference to a lexicon. "Any descriptive utterance, any statement in the true/false game" claimed Kuhn, "requires a prior lexicon, and that lexicon brings a sort of relativity with it" (1990b, p. 14). In other words, a lexicon makes possible any statement about the world and even truth itself. To clarify the position, he acknowledged that it resembles Kant's notion of categories in the *Critique of Pure Reason*, but, for Kuhn, the categories are "moveable." He then admitted that he requires Kant's *Ding an sich* to articulate adequately his position of reality. "It's the thing," claimed Kuhn, "about which nothing can be said but which legitimates what can be said properly" (1990b, p. 15). Thus, reality is not directly or absolutely knowable but rather it is "something" that simply makes possible knowledge about the world. Finally, Kuhn informed the audience that he has learned to live with this position and then inquired of it, "Am I realist, or am I not?" (1990b, p. 15).

Kuhn's question to the audience about whether he is a realist has generated considerable discussion and debate in the philosophical literature about his position with respect to realism, as well as to his idea of truth. Kuhn's critics have, notes Sankey, "detected a strong idealistic tendency in his views" (2000, p. 64). For example, Scheffler characterized Kuhn's position as "extravagant idealism." "Reality," lamented Scheffler over Kuhn's position, "is gone as an independent factor; each viewpoint creates its own reality. Paradigms, for Kuhn," he charged, "are not only 'constitutive of science'; there is a sense, he argues, 'in which they are constitutive of nature as well'" (1982, p. 19). However, in defense of Kuhn, Paul Hoyningen-Huene (1993) claims Kuhn's position is intermediate between the extremes of realism and idealism. Sankey also claims that Kuhn does not advocate a strictly mind-dependent idealism, and he argues that Kuhn's position is a form of constructive idealism, "which admits an independent reality but denies the possibility of epistemic access to it" (2000, p. 64). He goes on to identify Kuhn's position with Kant's position that the world is partly constituted through its conceptualization. Kuhn, for his part, denied that his position was idealistic or constructivistic (1991, p. 101). In the Thalheimer lectures, for example, he addressed the potential criticism that incommensurable lexicons and the world-changes associated with them represent a type of idealism. "Perhaps it is an idealist's world nonetheless," Kuhn admitted to the audience, "but it feels very real to me" (1984, p. 123). After all, stubbing one's toe hurts.

What then is Kuhn's stand on realism or the notion of reality? Given the latter switch to EPS and the maturation of IT_T, his position could be best described as adaptive realism. Even though the world is lexicon-dependent and thereby changes from one community to the next, Kuhn admitted it is "real" in the following sense.

> It provides the environment, the stage, for all individual and social life. On such life it places rigid constraints; continued existence depends on adaptation to them; and in the modern world scientific activity has become a primary tool for adaptation. What more can reasonably be asked of a real world? (1991, p. 10).

But as he quickly pointed out, the term "adaptation" is problematic because the process of adaption is reciprocal between a community and the world. Just as a species and its niche coevolve, so science and its world or niche also coevolve as the number of scientific specialties increases to carve the world up into narrower and narrower domains. "Conceptually," as Kuhn explained the position, "the world is *our* representation of *our* niche" (1991, p. 11). In other words, adaptive realism is a pragmatic notion in which a set of adaptive community practices does not progress towards closer approximations of what the world is, but rather it is a move away from a set of community practices that is not adaptive. Obviously, adaptive realism has important consequences for understanding truth.

As noted already, Kuhn rejected the correspondence theory of truth. Although he aimed to show that scientific practice is justified cognitively or rationally, Kuhn argued against "claims that successive scientific beliefs become more and more probable or better and better approximations to the truth" (1993, p. 330). Part of Kuhn's reasons for rejecting the correspondence theory was that a "stable Archimedean platform" for comparing successive paradigms is illusive since such a platform itself varies with changes in scientific beliefs (2000, p. 113). In addition, he invoked incommensurability, in terms of "no shared metric," which prohibits comparing successive articulations of the world because of the no-overlap principle. Kuhn offered the example of Aristotle's and Newton's notions of dynamics and asserted that the Newtonian lexicon "permits a more powerful and precise way than [Aristotle's] of dealing with what are *for us* the problems of dynamics, but these were not his problems, and lexicons are not, in any case, the sorts of things that can be true or false" (1993, p. 330). For Kuhn, Aristotle's statements about dynamics are not so much true or false *vis-à-vis* Newton's statements as they are "ineffable." Finally, justification of lexicons is not dependent on determining its truth or falsity, but rather it is pragmatic.

Critics have taken Kuhn to task for rejecting the correspondence theory of truth. Sankey, for example, asserts that Kuhn is "seriously mistaken" in rejecting the theory because, as he articulates Kuhn's position, "there is no basis on which to *judge* that theories are closer to the truth" (1998, p. 13). He argues that, because scientists articulate competing lexicons in a natural language, the truth of lexical statements can be compared to one another. According to Sankey, "the background natural language may serve as metalanguage for the lexicons, which may be treated as object-languages. Employing the natural language as metalanguage," he proposes, "it may then be said of some object-linguistic sentence from a given lexicon that it is true, while saying of another object-linguistic sentence from another lexicon that it is false" (1998, p. 13). Moreover, he objects to Kuhn's claim that truth is only intralexical since, according to Sankey, lexical terms are referents to natural kinds. As Sankey maintains, "there is a truth of the matter about the relation between lexicon and reality since a genuine question may be raised whether the entities postulated by a theory actually exist" (1998, p. 14). In other words, the truth of lexical statements can be assessed in terms of whether they correspond to entities within the world.

Sankey recognizes that Kuhn's justification for interlexical comparisons depends not on whether one lexicon converges on the truth but whether it is pragmatic. As he interprets Kuhn, "a lexicon has the status of a linguistic convention which may be judged on the basis of how well it serves a particular purpose rather than how well it reflects reality" (Sankey 1998, p. 14). What he fails to recognize about Kuhn's rejection of the correspondence theory and acceptance of the pragmatic justification of lexical change is that a lexicon is the outcome of an adaptive process between a community's practice and the world. In other words, the lexicon's "purpose" is not to articulate truth about the world but rather to formulate terms and statements that allow it to practice its trade successfully. In short, the point is survival not truth *per se*. Truth serves an instrumental function intralexically in weeding out inconsistent, incoherent, and contradictory statements so that communication among community members of a specialty remains open and is not impeded so progress can ensue. For Kuhn, scientific progress depends on open communication within a community. If incommensurability occurs, then communication may break down and the specialty may cease to progress or even become extinct.

Given this function for truth, Kuhn advocated replacing the correspondence theory with the redundancy theory of truth, also known as the disquotational theory of truth or no-truth theory (Schantz 2002). The redundancy theory is a minimalist or deflationary account of truth in which to claim a statement is true, is to claim its truth. For example, to assert, "snow is white," is to assert, snow is white. According to Kuhn, this theory would "introduce minimal laws of logic (in particular, the law of non-contradiction) and make adhering to them a precondition for the rationality of evaluations" (1991, pp. 8–9). Thus, truth plays simply an epistemic role to ensure proper assessment of empirical evidence and theoretical development—not an ontological role about what the world is substantively.

In summary, Kuhn's notions of truth and reality must be understood in the context of an instrumental or a pragmatic approach to scientific progress. Moreover, given the shift from HPS to EPS, Kuhn's notion of reality represents an adaptive realism as a community of specialists strives to develop its practice within a defined niche. Incommensurability plays a very important role, then, in a community's practice by isolating it from other communities so that its lexicon can be articulated without interference from them. This is especially the case for a community's lexicon that splits from a parent's lexicon. Kuhn's adaptive realism has important implications for the idea of truth. Truth is not some goal that science moves toward; rather, it represents an instrument for eliminating lexical statements that might contradict one another—i.e. it represents movement away from lexical statements that might prohibit progress. Hence, truth functions pragmatically to enhance the coherence of a community's lexicon so that communication among its members remains open so to promote progress. Finally, in rejecting the correspondence theory of truth, Kuhn advocated a redundancy or minimalist theory in which truth functions to ensure adherence of community members to a few logical laws such as non-contradiction.

9.7 Conclusion

In conclusion, although Kuhn remained committed to IT during his career, he modified it substantially in response to critics. Initially, IT_O represented the result of a "hermeneutical turn" for Kuhn in which he struggled to understand Aristotle's physics (see Fig. 9.1). He defined incommensurability as "no common meaning" between two successive paradigms. As such, it functioned to account for scientific progress during paradigm shifts or scientific revolutions. Kuhn's shift from IT_O to IT_P was the consequence of a "linguistic turn" in which he abandoned the visual metaphor of a gestalt switch for one of linguistic communities. He also abandoned paradigm for disciplinary matrix and exemplar, and then defined incommensurability as "no common values and exemplars." However, the role of IT_P remained similar to the role for IT_O. Kuhn's shift from IT_P to IT_L was the outcome of further development in the "linguistic turn" in which he substituted the notion of a scientific community's lexicon and its taxonomic categories for the concepts of disciplinary matrix and exemplars. Kuhn now defined incommensurability as "no common language;" but, IT_L continued to function similarly to IT_O and IT_P.

Kuhn's final shift from IT_L to IT_T represents an "evolutionary turn," in which he substituted an EPS for a HPS. He defined incommensurability in terms of "no common taxonomy" and added taxonomic hierarchies to the lexicon. Incommensurability's role now involved the isolation of a new lexicon to permit its development. Lastly, Kuhn's development of IT_T had important implications for an approach to reality and truth. As for reality, Kuhn's position can be described as adaptive realism in which a community of specialists strives to develop its practice within a defined niche. As for truth, it functions pragmatically as an instrument for eliminating lexical statements that might contradict one another and thereby hinder lexical development and ultimately scientific progress in terms of increased specialties. Hence, truth enhances the coherence of a community's lexicon so that communication among its members remains open so to promote progress.

References

Barker, P. 2001. Kuhn, incommensurability, and cognitive science. *Perspectives on Science* 9:433–462.
Bird, A. 2011. Thomas Kuhn's relativistic legacy. In *A companion to relativism,* ed. S. D. Hales, 475–488. Oxford: Blackwell.
Davidson, D. 1974. On the very idea of a conceptual scheme. *Proceedings and Addresses of the American Philosophical Association* 47:5–20.

Gattei, S. 2008. *Thomas Kuhn's "linguistic turn" and the legacy of logical positivism: incommensurability, rationality and the search for truth.* Burlington: Ashgate.

Hoyningen-Huene, P. 1993. *Reconstructing scientific revolutions: Thomas S. Kuhn's philosophy of science.* Chicago: University of Chicago Press.

Hoyningen-Huene, P. 2005. Three biographies: Kuhn, Feyerabend, and incommensurability. In *Rhetoric and incommensurability,* ed. R. A. Harris, 150–175. West Lafayette: Parlor Press.

Hung, E. H.-C. 2006. *Beyond Kuhn: Scientific explanation, theory structure, incommensurability and physical necessity.* Burlington: Ashgate.

Kuhn Papers, and S. Thomas. Massachusetts Institute of Technology, Institute Archives and Special Collections. Cambridge, Massachusetts.

Kuhn, T. S. 1957. *The Copernican revolution: Planetary astronomy in the development of western thought.* Cambridge: Harvard University Press.

Kuhn, T. S. 1962. *The structure of scientific revolutions.* Chicago: University of Chicago Press.

Kuhn, T. S. 1970. *The structure of scientific revolutions.* 2nd ed. Chicago: University of Chicago Press.

Kuhn, T. S. 1983. Commensurability, comparability, communicability. *Philosophy of Science Association* 1982 (2): 669–688.

Kuhn, T. S. 1984. *Thalheimer lectures: Scientific development and lexical change.* Kuhn Papers, MC 240, boxes 21–22.

Kuhn, T. S. 1987. Sherman lectures: The presence of past science. Kuhn Papers, MC 240, boxes 30–32.

Kuhn, T. S. 1989a. Possible worlds in history of science. In *Possible worlds in humanities, arts and sciences,* ed. S. Allén, 9–32. Berlin: Walter de Gruyter.

Kuhn, T. S. 1989b. NSF Application. Kuhn Papers, MC 240, box 20.

Kuhn, T. S. 1990a. Dubbing and redubbing: The vulnerability of rigid designation. In *Scientific theories,* ed. C. W. Savage, 298–318. Minneapolis: University of Minnesota Press.

Kuhn, T. S. 1990b. UCLA cognitive science colloquium. Kuhn Papers, MC 240, box 24.

Kuhn, T. S. 1991. The road since structure. *Philosophy of Science Association 1990* (2): 3–13.

Kuhn, T. S. 1993. Afterwords. In *World changes: Thomas Kuhn and the nature of science,* ed. P. Horwich, 311–341. Cambridge: MIT Press.

Kuhn, T. S. 2000. The trouble with the historical philosophy of science. In *The road since structure. Philosophical essays, 1970–1993,* ed. J. Conant and J. Haugeland, 105–120. Chicago: University of Chicago Press.

Lakatos, I. 1970. Falsification and the methodology of scientific research programmes. In *Criticism and the growth of knowledge,* ed. I. Lakatos and A. Musgrave, 91–196. Cambridge: Cambridge University Press.

Marcum, J. A. 2013. Whither Kuhn's historical philosophy of science? An evolutionary turn. In *An anthology of philosophical studies,* ed. P. Hanna, vol. 7, 99–109. Athens: Athens Institute for Education and Research.

Musgrave, A. E. 1971. Kuhn's second thoughts. *British Journal for the Philosophy of Science* 22:287–306.

Popper, K. R. 1970. Normal science and its dangers. In *Criticism and the growth of knowledge,* ed. I. Lakatos and A. Musgrave, 51–58. Cambridge: Cambridge University Press.

Sankey, H. 1998. Taxonomic incommensurability. *International Studies in the Philosophy of Science* 12:7–16.

Sankey, H. 2000. Kuhn's ontological relativism. *Science & Education* 9:59–75.

Sankey, H., and P. Hoyningen-Huene. 2001. Introduction. In *Incommensurability and related matters,* ed. P. Hoyningen-Huene and H. Sankey, vii–xxxiv. Dordrecht: Kluwer.

Schantz, R., ed. 2002. *What is truth?* New York: Walter de Gruyter.

Scheffler, I. 1982. *Science and subjectivity.* 2nd ed. Indianapolis: Hackett.

Shapere, D. 1964. The structure of scientific revolutions. *Philosophical Review* 73:383–394.

Shapere, D. 1971. The paradigm concept. *Science* 172:706–709.

Siegel, H. 1987. *Relativism refuted: A critique of contemporary epistemological relativism.* Dordrecht: Kluwer.

Sigurdsson, S. 1990. The nature of scientific knowledge: An interview with Thomas Kuhn. *Harvard Science Review* 3:18–25.

Wang, X. 2007. *Incommensurability and cross-language communication.* Burlington: Ashgate.

Wray, K. B. 2011. *Kuhn's evolutionary social epistemology.* Cambridge: Cambridge University Press.

Chapter 10
Walking the Line: Kuhn Between Realism and Relativism

Michela Massimi

10.1 Introduction

Fifty years after the publication of *The Structure of Scientific Revolutions* (henceforth SSR), Kuhn's view continues to exercise a never-ending fascination among historically inclined philosophers of science. There is more to Kuhn's fascination among acolytes than the recent trend of "integrated history and philosophy of science"; or, simple tribute to one of the most influential figures of the past century. For a generation like mine, who grew up and was trained in philosophy of science, when debates about Kuhn were rapidly being replaced by the latest additions (e.g., constructive empiricism, model-based accounts of science, and so forth), Kuhn represents the main advocate of a philosophical tradition that has seriously challenged the realist credo. Was Kuhn some kind of relativist? Did he advocate a form of constructivism? Or was he a mild realist, after all?

Kuhn's take on scientific realism is well known. His attack on the positivist, cumulative model of knowledge acquisition has often been read as committing him to scientific anti-realism. His rejection of truth as correspondence with well-defined, cross-paradigm facts qualifies him as a metaphysical anti-realist. But can Kuhn be regarded as a realist (with some suitable caveats)? Among the many available readings of Kuhn, two have portrayed Kuhn as a mild, sophisticated realist. The first is Paul Hoyningen-Huene's (1993, and this volume, Chap. 13) Kantian reading. The second is Ron Giere's (2013) more recent "perspectival realist" reading of the late Kuhn (i.e., the Kuhn of *The Road Since Structure*, whereby paradigms are replaced by scientific lexicons).

I concentrate here on these two readings. I review them briefly, and focus on some of the challenges facing them. In Sect. 10.2, I discuss Hoyningen-Huene's Kantian

M. Massimi (✉)
School of Philosophy, Psychology and Language Sciences,
University of Edinburgh, Edinburgh, Scotland, UK
e-mail: michela.massimi@ed.ac.uk

© Springer International Publishing Switzerland 2015   135
W. J. Devlin, A. Bokulich (eds.), *Kuhn's Structure of Scientific Revolutions—50 Years On*,
Boston Studies in the Philosophy and History of Science 311,
DOI 10.1007/978-3-319-13383-6_10

reading and Bird's two challenges against it: what I call the *challenge of naturalism* and *the challenge of phenomena*. In Sect. 10.3, I turn my attention to Ron Giere's perspectival realist reading, and what I take to be its two main problems: *conceptual relativism and alethic relativism*. In Sect. 10.4, I introduce the notion of *naturalized Kantian kinds* (or NKKs). I clarify what NKKs are, and how they blend the Kantian reading of Kuhn with a Quinean naturalistic reading of kinds. Finally, in Sect. 10.5, I put NKKs to use by showing how they can deliver a mildly realist reading of Kuhn, which eschews some of the challenges facing the other two readings.

As a ground-clearing remark, I do not mean to suggest that Kuhn ever endorsed or even contemplated the possibility of what I call NKKs—as such, my line of argument is neither interpretive nor exegetical. I have argued elsewhere for what I take to be Kuhn's own view about world-changes as a semantic, rather than a metaphysical thesis (Massimi, 2014b). Hence, the present paper should not be read as my way of interpreting Kuhn's own view. Instead, I am here engaged in a different exercise: namely, to understand what realist metaphysics *might* be available to vindicate Kuhn's intuitions concerning paradigm shifts and world-changes. I suggest that NKKs provide just the realist metaphysics we need. It may not be strong enough to qualify the end product as a form of scientific realism (certainly, it is not a form of metaphysical realism). But it is resilient enough to avoid the blanket charges of anti-realism and relativism; or so I shall argue.

10.2   The Kantian Reading of Kuhn

In 1993, Hoyningen-Huene brought to the scene an ifluential Kantian reading of Kuhn, which met with Kuhn's own approval (Kuhn had been investigating himself the Kantian roots of his view as early as 1979, see: Kuhn 1991, p. 12; Kuhn 2000, p. 245). Key to this reading is the identification of two different meanings for "world" or "nature" in the context of Kuhn's much discussed claim that after a scientific revolution scientists live in a new world (Kuhn 1962, p. 121). Hoyningen-Huene has argued that "world" should be identified with what Kant would call the world of *phenomena* or *objects of experience*. This is the world that we can have knowledge of, and it is "determined jointly by nature and the paradigms" (Kuhn 1962, p. 125). It is a world of spatiotemporal appearances, which have been conceptualized according to the dominant paradigm. But "world" may also denote the world of the things-in-themselves, which Kant deemed unknowable. On Hoyningen-Huene's reading, Kuhn's contentious claim about world-changes should be read in light of this twofold meaning for "world". As such, instead of a claim of dubious constructivist flavor, it becomes a claim about the phenomenal world as we come to know it (see also Hoyningen-Huene 2008, p. 44, fn. 2). Obviously, the Kantian reading needs be updated in the light of history of science. For Kant, the forms of sensibility and a priori categories of the understanding were universal and fixed once and for all. For Kuhn, the scientific paradigms (and later, the lexicons) provide the conditions of possibility of knowledge; and these conditions change over time and after a scientific revolution. Thus, Peter Lipton's apt expression of Kuhn as "Kant on Wheels" (Lipton 2001).

How persuasive is this Kantian reading? Critics have pointed out that it is at odds with Kuhn's naturalism (Bird 2012, p. 869). Kuhn was deeply influenced by contemporary work in psychology and cognitive sciences, as the numerous references to leading gestalt psychologists in SSR clearly show. Much of his discussion about paradigm shift and incommensurability can be read in a naturalistic vein (see also Bird 2000, and this volume, Chap. 3)—as emanating from his life-long engagement with the cognitive sciences, following a trend opened by N.R. Hanson's *Pattern of Scientific Discovery* (1958). If the naturalistic take on Kuhn is correct, it poses a challenge to the Kantian reading. How can the phenomenal world, and the world of things-in-themselves, square with gestalt psychology and duck-rabbit images? Let us call this the *challenge of naturalism* (CoN):

> (CoN) The "world" of scientific inquiry (with its objects) is the product of our causal interaction with the physical world (via sense data, stimuli, background beliefs and so forth).

It is easy to see why CoN poses a problem for the Kantian reading of Kuhn. If we identify the physical world in the definition above with the second aforementioned meaning of "world", qua noumenal world of things-in-themselves, then we are in no position to explain how the world of scientific inquiry could possibly ensue from it (as CoN would have it). For the noumenal world is unknowable, nor is it amenable to causal interactions. We do not *causally interact* with things-in-themselves when we 'see' the world differently, because things-in-themselves—by definition —do not enter into a two-place causal relation with us qua cognizing agents. Thus, if the naturalistic take on Kuhn is correct (as textual evidence would suggest), it is bad news for the Kantian reading.

A possible line of response on behalf of the Kantian reading would be to insist that things-in-themselves do *cause* after all the phenomena that we come to know. Even within the resources of Kant's own view, there may be ways of accommodating this intuition (see Langton 1998; Chignell 2010). I won't pursue this line of response here, as it would lead me astray from my purpose. Hoyningen-Huene (1993, 43 ff.) has addressed CoN by showing how especially after SSR, Kuhn defended a form of stimulus ontology: scientific revolutions bring about different worlds by presenting different communities with different data by the same stimuli (Kuhn 1977, p. 309). If we replace things-in-themselves with a posited world of stimuli, we could reconcile CoN with the Kantian reading. For stimuli relate to sensations, like noumena relate to phenomena. Like noumena, stimuli are not 'given' to us. Yet, by contrast with noumena, stimuli (be they sound waves, photons, or other) can be investigated by empirical science and are *causally efficacious* in producing sensations about the external world.

Let us grant for the sake of the argument that a shift to stimulus ontology may reconcile the Kantian reading with the naturalistic stance recommended by CoN. More problems loom on the horizon. If stimuli are not 'given' to us and hence are indescribable (like the noumena), how can we tell apart *equivalent* stimuli from *different* ones? If stimuli *cause* the sensations/phenomena, presumably different phenomena are caused by different stimuli. But we *cannot know* whether different stimuli do indeed cause different phenomena, given the epistemic non-accessibility of stimulus ontology. Even more problematic is Kuhn's occasional slip of the pen about whether Aristotle and Galileo "really *see* different things when *looking at* the same sort of objects?" (1962, p. 120). For it assumes that the same stimulus/thing-in-itself can be *looked at* in different ways by different scientists.

Yet, Kuhn quickly rectified the slip of the pen: the whole idea that Aristotle and Galileo simply differed in their *interpretation* of the pendulum is flawed and untenable, according to Kuhn. It is the product of a philosophical tradition (beginning with Descartes and continuing with Newton), which "served both science and philosophy well", but has proved false by new research in "philosophy, psychology, linguistics, and even art history" (1962, p. 121). For, "pendulums were brought into existence by something very like a paradigm-induced gestalt switch" (1962, p. 120).

The problem remains. How can *equivalent* stimuli (e.g. looking at the pendulum) lead to gestalt-switches and different phenomenal worlds in a Kantian sense? As Hoyningen-Huene (1993, pp. 50–60) acknowledges, attempts to understand Kuhn's view in the light of stimulus ontology are fraught with difficulties. In trying to square the Kantian reading with CoN via stimulus ontology, we end up with another puzzle, which Bird (2012, p. 870) captures well when he objects that "it is not especially plausible to say that Aristotle and Galileo have different visual experiences when looking at a pendulum, and it is even less plausible to think in terms of changes in sensory experience when we turn to the relativistic revolution". I call this the *challenge of phenomena* (CoP):

> (CoP) If world-changes are changes in *phenomenal worlds* (along Kantian lines), they cannot just be changes in visual and sensory experiences. *Phenomenal worlds* are not reducible to psychological gestalt switches.

The two challenges (CoN and CoP) pull in opposite directions. On the one hand, to meet the challenge of naturalism, the Kantian reading has to engage with stimulus ontology and explain world-changes along the lines of gestalt psychology. On the other hand, by doing so, the Kantian reading gets robbed of its distinctive character, because there is more to Kantian phenomena than psychological gestalt switches. At stake here is the notion of Kantian phenomena, which proves too strong to be reconcilable with Kuhn's naturalism (as CoN would have it).

Where does this discussion leave us? The Kantian reading has a lot going for it, not least the ability to capture the metaphysical anti-realist spirit of Kuhn's enterprise. Yet the identification of Kuhn's "worlds" with Kant's phenomena require further elaboration along the lines of naturalism, but beyond psychological gestalt-switches. I turn to this task in Sect 10.4. Before that, I want to mention another recent realist reading of Kuhn by Ron Giere—this time as a "perspectival realist". Interestingly, there are some common features between these two realist readings (the Kantian and the perspectival realist). Both vindicate the human vantage point against any God's eye view, so to speak. Each achieves this task by placing the metaphysical burden on either a Kantian notion of phenomena, or a historically-defined scientific perspectives.

10.3   The Perspectival Realist Reading of Kuhn

With the publication of *Scientific Perspectivism* (2006), Giere has launched a new trend that promises to do justice to Kuhn's view of scientific knowledge, while avoiding the excesses of both 'objectivist' realism and relativism. Perspectival realism promises to go beyond "The Science Wars" and to offer a much-needed rapprochement between science studies and philosophy of science. Or, between those among us, who embraced the Kuhnian lesson and went for a sociologically informed philosophy of science; and those who, instead, held back onto notions of epistemic warrant and scientific rationality. The philosophical pedigree of Giere's scientific perspectivism stretches back to Leibniz, Kant, and Nietzche (Giere 2006, p. 3). And it shares with Hilary Putnam's internal realism, the rejection of any God's eye view on nature. Its manifesto reads as follows: "According to this highly confirmed theory (or reliable instrument), the world seems to be roughly such and such'. There is no way legitimately to take the further objectivist step and declare unconditionally: 'This theory (or instrument) provides us with a complete and literally correct picture of the world itself'" (Giere 2006, p. 6). The view is fully naturalistic in rejecting a priori claims of any kind and in deferring to the empirical sciences (and in particular, to a model-based view of how science works, whereby scientific representation involves agents using models to achieve specific goals).

This is realism insofar as it implies the belief in a world "out there", which is the object of our representational practices (no matter from which vantage point, or perspective, the representation may take place). Thus, one may say that the White House is off to the right of the Washington Monument, if looked at from the steps of the U.S. Capital Building. But it would be off to the left of the Washington Monument, if looked at from the steps of the Lincoln Memorial (Giere 2006, p. 81). Our chosen vantage point does not affect the reality of the White House or the Washington Memorial; instead, it affects only what we can legitimately say about their reciprocal spatial relations.

Yet things are not as simple as the perspectival metaphor may suggest. When we utter a sentence such as: "The White House is off to the right of the Washington Monument", we want to know whether the proposition expressed is true or false. The perspectival realist would reply that whether it is true or false depends on our perspective (i.e. whether we are standing on the steps of the Lincoln Memorial or on the steps of the U.S. Capital Building). That is what makes Giere's position "perspectival", after all: "For a perspectivist, truth claims are always relative to a perspective" (2006, p. 81). Hence, the similarity with Kuhn's view: "Claims about the truth of scientific statements or the fit of models to the world are made within paradigms or perspectives" (2006, p. 82). Despite the similarity with Kuhn's view, Giere warns us that scientific perspectives cannot however be identified *tout court* with paradigms. Scientific perspectives are narrower than a disciplinary matrix, which would include beliefs, values and techniques shared by a community at a given time. But they are also broader than Kuhn's exemplars (2006, p. 82), intended as a paradigmatic application of a particular type of representational model (say, the pendulum).

One advantage of Giere's view is that it seems to avoid the linguistic incommensurability of the late Kuhn. No problem of translation arises in between scientific perspectives, precisely because they are narrower than disciplinary matrices. There might be specific problems in comparing different scientific perspectives; for example, in comparing measurement outcomes of different scientific instruments (say PET and fMRI scans of the brain—2006, p. 83). But these are problems with well-known solutions. They do not open the door to incommensurable scenarios in Kuhn's sense.

The other advantage of Giere's perspectivism (this time over scientific realism) is that it seems to deliver a notion of "natural kinds" that is malleable and flexible enough to avoid the unpalatable consequences of taxonomic monism. The case for perspectival realism seems to find support in biology, where the species problem clearly demonstrates the limits of any objectivist take on biological kinds and speaks in favor of a perspectival reading. What counts as a 'species' is ultimately functional to what theoretical perspective (e.g., evolutionary taxonomy or cladistics) is endorsed. A similar point can be made about chemical kinds. By drawing on Joseph LaPorte's recent take on Putnam's Twin Earth story (LaPorte 2004, Chap. 4), Giere notes (2006, p. 86) how there is no fact of the matter about classifying chemical kinds on the basis of atomic number as opposed to atomic weight. There are properties such as freezing and melting points, or chemical reactivity that vary significantly among isotopes of the same element. Thus, a perspectival reading can do justice to biological kinds and chemical kinds better than any objectivist readings.

Was Kuhn himself a perspectival realist? Or, at least, can we understand Kuhn's philosophical enterprise along perspectivalist lines? This is what Giere suggests in a recent paper (Giere 2013). By concentrating primarily on the late Kuhn of *The Road Since Structure* (Kuhn 2000), whereby lexicons replace paradigms and Newtonian mechanics or Cartesian physics can be regarded as scientific perspectives, Giere distinguishes between two possible versions of perspectivism.

The first version of perspectivism at work in Kuhn has to do with the many possible ways of *classifying* the same objects. While today we still share with ancient Greek astronomers the names of many celestial bodies (Earth, Moon, Sun), we nonetheless classify them differently (e.g., the Sun is regarded as a star, not as a planet). Hence, the incommensurability between the Ptolemaic and the Copernican lexicons. Despite sharing the same kind terms (e.g., 'planet') and even the same names (e.g. 'Sun'), sentences in the Copernican lexicon cannot be translated into the Ptolemaic lexicon because an important change has occurred in the taxonomic categories at issue, so that the two terms (e.g., 'planet' pre-Copernicus, and post-Copernicus) overlap in the two lexicons without a complete overlapping of their respective extensions. This is the way in which the late Kuhn redefined incommensurability as untranslatability between scientific lexicons (Kuhn 2000, pp. 92–93). Giere observes that we should understand "physical categories as defining perspectives within which to interpret the physical world. In short, Kuhn seems to be a scientific perspectivist regarding all the sciences, natural as well as social" (Giere 2013, p. 54). Thus, the first version of perspectivism in Kuhn (let us call it PiKu$_1$) can be summarized as follows:

(PiKu$_1$): Scientific perspectives are defined by the taxonomic categories (both natural and social) of a scientific lexicon, through which we *interpret* the physical world.

Thinking of scientific perspectives as lenses, through which we can interpret the physical world seems in line with the late Kuhn's take on W.V.O. Quine's radical translator as an *interpreter* (see Kuhn 2000, 37 ff.). The ontological relativity ensuing from Quine's argument seems also to resonate with Kuhn's claim about world-changes. Thus, on a first possible perspectivist reading of Kuhn, scientific perspectives are conceptual taxonomies, through which we *classify and interpret* the world.

The second version of perspectivism that can be found in the late Kuhn is more Kantian than Quinean in spirit. Under the influence of Michael Friedman's reading of Kant and Reichenbach, Kuhn came to refer to his own view of scientific lexicons as resembling Kant's a priori in a relativized form (Kuhn 2000, p. 245; see also Kuhn 1993, pp. 331–332). This provides the springboard for Giere's second perspectivist reading of Kuhn: "I would assimilate a relativized, thus contingent, set of constitutive principles as defining a theoretical perspective, within which one could formulate potentially true statements. This is a version of perspectivism" (Giere 2013, p. 54).

On this second reading, the emphasis shifts from conceptual taxonomies to what Friedman would call (using Kant's terminology) *constitutive principles*, principles that are constitutive for the possibility of our experience of the world. In a Newtonian world, such principles would be Newton's three laws of motion, for example; in Einstein's world, they would be the light principle of special relativity and the equivalence principle of general relativity. These principles are *constitutive a priori* in a somehow Kantian sense, because they must be in place for our experience of the physical world (qua a Newtonian or an Einsteinian world) to be possible at all. In other words, they are constitutive because they are necessarily presupposed for the proper empirical part of a theory (e.g. Newton's law of gravity in Newtonian mechanics, or Einstein's field equations in general relativity) to be possible (see Friedman 2001). At stitutive a priori principles are revisable and do change after a scientific revolution, as indeed the transition from Newtonian mechanics to Einstein's relativity theory shows. Hence they are *relativized a priori* as Reichenbach conceived of them. Thus, this second version of perspectivism in Kuhn (let us call it PiKu$_2$) can be summarized as follows:

(PiKu$_2$): Scientific perspectives are defined by the theoretical principles of a scientific lexicon, through which we can *experience* the physical world.

This second version of perspectivism is stronger than the first one in taking scientific perspectives as more than conceptual taxonomies to interpret the world. Scientific perspectives incorporate relativized constitutive principles, qua *conditions of possibility of our experience* of the world. Constitutive a priori principles (say, Newton's three laws of motion in Newtonian mechanics) provide the conditions of possibility of what we can (truly or falsely) assert about objects in motion within this scientific perspective. For example, starting with the concept of inertial mass and Newton's second law as a constitutive a priori principle, we can introduce Newton's law of gravity and the related concept of weight to make empirical judgments about motions ensuing from gravitational attraction between the Sun and the Earth. Key to this second perspectivist reading of Kuhn are laws of nature (as opposed to taxonomic categories) built into the lexicon as "synthetic a priori" (see Kuhn 1990, p. 306).

Giere's overall reading of Kuhn as a perspectival realist is on the right track, in my opinion, and is corroborated by textual evidence. Yet, like Hoyningen-Huene's Kantian reading, Giere's reading too has to confront two difficulties. How can Giere's reading deliver on the promise of realism, even of a 'perspectival' kind? The first difficulty concerns $PiKu_1$. If scientific perspectives are indeed identified with taxonomic categories through which we interpret the world, it is hard to escape the conclusion that there is no fact of the matter about the world, *independently of* the taxonomic categories through which we 'see' the world. When our taxonomies change (after a scientific revolution), the world itself changes because the phenomena we have access to are *categorized* and *classified* differently. $PiKu_1$ then addresses the *challenge of phenomena* (CoP) described in Sect. 10.2 above, by providing a more robust reading of phenomena than mere visual and sensory gestalt switches. There is, however, a price to pay for this more robust reading of phenomena: the kind of Kantianism at work in $PiKu_1$ resembles dangerously a form of conceptual relativism. After all, was not it Kant, who introduced the distinction between form and content, between the categories of the understanding and the empirical manifold given to us via sensibility? Aren't Kantian phenomena the end products of our categories and concepts acting as cookie-cutters on the worldly dough? If scientific perspectives are identified with taxonomic categories (as per $PiKu_1$), we can read Kuhn's claim about world-changes as a claim about changes in phenomenal worlds (as Hoyningen-Huene would have it) but at the cost of identifying Kantian phenomena with *perspective-dependent conceptualized appearances*. Hence, the ensuing challenge of conceptual relativism (CR) :

(CR): Under $PiKu_1$ reading, changes in the phenomenal worlds are due to changes in the taxonomic categories. Different taxonomic categories produce different phenomena qua *perspective-dependent conceptualized appearances*.

Thus, Aristotle and Galileo 'worked' in different worlds, because where Aristotle saw accelerated motion towards a natural place for a free falling stone, Galileo saw accelerated motion from the origin due to a weight-related concept of 'gravitas'. Scientific perspectives become conceptual frameworks, beyond which there is no ready-made or perspective-independent world. This may seem a rather innocuous claim, like the claim that there is no perspective-independent view of the White House. But it has in fact a hefty metaphysical price. If Kuhn's view deserves the title of realism (even of a perspectival kind), it ought to be possible for Kuhn's 'worlds' to be perspective-independent. There have to be perspective-independent facts or states of affairs, not molded by our taxonomic categories. But $PiKu_1$ precludes this possibility. Our interpretation of the world is always from the vantage point of a conceptual taxonomy or another. There is no view from nowhere, and no phenomenal world without a conceptual taxonomy in place. Following PiKu$_1$, we are left with a form of conceptual relativism about what there is.[1]

Are the prospects of PiKu$_2$ any more promising? What if we take Kuhn's 'phenomenal worlds' as the expression of different constitutive principles at work in different lexicons? Shifting the focus from taxonomic categories to laws of nature and theoretical principles has the obvious advantage of avoiding the relativism that we saw displayed by (CR). It would be possible to reconcile the perspectivalist reading with Hoyningen-Huene's Kantian reading of Kuhn, this time by placing the burden of proof on principles *constitutive of our experience of the world*. Aristotle and Galileo 'worked' in different worlds, because where Aristotle saw accelerated motion towards a natural place for a free falling stone, Galileo (and Newton after him) saw accelerated motion as an instantiation of what would later be described as Newton's second law, (i.e. as motion due to an accelerating force). Scientists working with the Newtonian perspective/ lexicon learn how to *experience the world* along the lines of Newton's laws with all its exemplars. PiKu$_2$ addresses the *challenge of phenomena* (CoP) described in Sect. 10.2, by providing a more robust reading of phenomena than mere visual and sensory gestalt switches, yet without landing us into conceptual relativism. Thus, it would appear that the prospects for realism are better under PiKu$_2$.

But, also in this case, a difficulty looms on the horizon. If Kuhn's view deserves the name of realism, not only must there be perspective-independent facts or states of affairs. What we can also *truly* assert about those facts should not depend on our scientific perspective. Truth and falsity cannot be perspective-dependent, this time

---

[1] The problem does not arise if one is willing to endorse or entertain the possibility of conceptual relativism, of course. Appealing as conceptual relativism might be as a philosophical position, the point I want to make here is a different one. Namely, that if Kuhn's view has to be qualified as a form of realism, it cannot possibly fall into the traps of conceptual relativism. Kuhn's view is either relativist or realist; it cannot be both at the same time. One may respond at this point that PiKu1 does not, after all, rule out perspective-independent facts. We may, for example, assume that there are natural kinds in nature, which none of our conceptual taxonomies gets exactly right (because the world is too complicated). But some conceptual taxonomies get closer and in so doing, they tell us a lot about the world. Thus, on this reading, PiKu1 is compatible with there being perspective-independent facts (I thank Paul Teller for this comment). In reply, it is worth noting that this might be (or may be intended to be) a possible reading of the Kantian relation between the noumenal world and the phenomenal world. But it does not work as a reading of Kuhn (even as a Kantian-flavoured perspectival reading of Kuhn). For it takes Kuhnian conceptual taxonomies as representational schemes that do not get the world exactly right. But Kuhn never understood conceptual taxonomies as sheer representational schemes for a mind-independent world. On the contrary, he entrusted conceptual taxonomies with the role of opening up entire worlds by affecting the very same experimental data, measurement techniques, nomic generalisations, and ensuing classifications of objects into natural kinds.

on pain of *alethic relativism* (AR), i.e. what is true in one scientific perspective is false in another:[2]

> (AR): Under PiKu$_2$ reading, changes in phenomenal worlds are due to changes in the constitutive theoretical principles of a lexicon. Different theoretical principles make possible different *experiences of the world* and of what we can truly assert about it.

Giere acknowledges the problem and quotes Kuhn saying that "Each lexicon makes possible a corresponding form of life within which the truth or falsity of propositions may be both claimed and rationally justified, but the justification of lexicons or of lexical change can only be pragmatic" (Kuhn 2000, p. 244; quoted in Giere 2013, p. 54). Giere himself seems to accept alethic relativism by identifying lexicons with perspectives, within which to formulate truth claims (2013, p. 55). As announced at the beginning of this paper, my goal here is not exegetical; as such, I do not want to question whether or not Kuhn might have been close to endorsing some form of alethic relativism. But one thing is clear. If Kuhn is a perspectivalist along the lines of PiKu$_2$, he cannot be legitimately called a 'realist' too (on pain of perspectival realism being compatible with a form of alethic relativism). Realism—I take it—is incompatible with both facts and truths being relative to incommensurable scientific perspectives. Relativism about facts and relativism about truths are at odds with the metaphysical and the epistemic tenets of realism, respectively. Neither PiKu$_1$ nor PiKu$_2$ seem to deliver on the promise of realism.

Let us take stock. In Sect 10.2, I discussed Hoyningen-Huene's Kantian reading of Kuhn and two main challenges (the challenge of naturalism, CoN, and the challenge of phenomena, CoP). In Sect 10.3, we considered Giere's perspectival realist take on Kuhn. The challenge of phenomena can be addressed by interpreting Kuhn's phenomena along the lines of two possible perspectivalist readings (PiKu$_1$ and PiKu$_2$). We assessed the promise and prospects of both readings for delivering a *realist* reading of Kuhn and concluded that both face difficulties. Where to go next? In the next section, I take up the challenge of defending a perspectival realist reading of Kuhn. This perspectival realist reading can do justice to Hoyningen-Huene's Kantian take on Kuhn and address the challenge of phenomena, without yet committing us to either conceptual relativism or alethic relativism. Central to it is the notion of naturalized Kantian kinds (NKKs), to which I now turn.

---

[2] One may reply that truth and falsity are indeed perspective-dependent. Teller (2011), for example, argues that both sentences "water is a continuous fluid" and "water is a statistical collection of molecules" are true in their respective perspectives (i.e. hydrodynamics and statistical mechanics), despite being in conflict with one another. If we understand again perspectives as 'idealised representational schemes', no threat of alethic relativism ensues. In reply to this point, I concede that Teller's way of characterising scientific perspectives avoids the risk of alethic relativism, and, I would argue, it is in fact closer to a form of contextualism than relativism. But again one may wonder whether Teller's characterisation of perspectives captures the spirit of Kuhn's view, and whether Giere's characterisation comes instead closer. It is to Giere's reading of Kuhn as a perspectivalist that I focus on here.

10.4   Naturalised Kantian Kinds (NKKs)

Think of scientific kinds, i.e. the kinds of entities and objects described or hypothesized by scientific theories: phlogiston, oxygen, ether, electromagnetic field, but also neutrinos, bacteria, microbes, DNA, and so forth. Are these kinds *natural kinds*? Realism would have it that some of them are (e.g. oxygen, electromagnetic field, neutrinos, etc.), but others clearly are not (e.g. phlogiston, ether, etc.). How do we tell apart the scientific kinds which are natural from those which are not? According to realism, success is key to this process. The natural kinds are those kinds described by *successful scientific theories*, those theories that have proved resilient to falsification, have delivered novel predictions, and successfully latch onto the causal structure of the world; or so scientific realism claims.

Kuhn's argument for incommensurability and world-changes has traditionally been a powerful counterargument to this realist claim. For Kuhn showed that success is relative to the scientific paradigm/ lexicon in place. More to the point, Kuhn proved that the realist belief in the ability of our scientific theories to causally track natural kinds is bankrupt. Kuhn famously rejected the Kripke-Putnam view of natural kind terms as rigid designators (Kuhn 1990). Against Putnam's causal theory of reference, Kuhn argued that it is not the case that the term "water" rigidly designates the same kind of stuff before and after Lavoisier's Chemical Revolution (i.e. before and after the discovery that water is H_2O). Back in 1750, before the Chemical Revolution, states of aggregation determined chemical species, and liquidity was regarded as an essential property of water. It was one of the achievements of the Chemical Revolution to take states of aggregation as marking physical, rather than chemical, species. Thus, back in 1750 "water" referred to *liquid* H_2O. Kuhn concluded that whether properties are essential or accidental in defining a natural kind is contingent on the scientific paradigm, and Putnam's causal theory of reference leaves unscathed the problem of meaning-change.

Kuhn reasserted the same point in his reply to Hacking's (1993) nominalist take on kinds in his proposed solution to the problem of world-changes. This time, Kuhn argued that it is difficult to understand along *nominalistic* lines the referents of terms such as "force" and "wave front" (pace Hacking), and a notion of "kind" was needed to populate the world and to divide pre-existing populations (Kuhn 1993, p. 319). The problem—as Kuhn saw it—is that terms such as "water$_1$" and "water$_2$" (before and after the Chemical Revolution; or before and after the discovery of isotopes) are projectable kind terms, embodying different expectations. Hence, although the same term is seemingly applied to the same object, substantial changes in the underlying meaning and nomic expectations have occurred in the meantime. Natural kind terms are not rigid designators, and nothing warrants the realist belief that our successful theories causally track natural kinds.

Scientific kinds are then central to Kuhn's view about incommensurability and world-changes. His anti-essentialist and anti-nominalist take on kinds is pivotal to his claim that scientists live and work in different worlds after a revolution. Worlds are made of *kinds* (e.g. kinds of entities, kinds of forces, and so forth). But kinds neither carve nature at its joints, nor are conventional bundles of individuals. Thus, the question naturally arises: What are kinds? How should we understand the notion of scientific kinds in Kuhn?

In what follows, I propose that we understand scientific kinds as *naturalized Kantian kinds* (or NKKs). I am not suggesting that Kuhn endorsed this view or even possibly dreamt of it. Instead, I suggest that *we interpret* Kuhn's aforementioned remarks along these lines. Why? Because we might have in our hands a way of understanding Kuhn's world-changes along Kantian lines (as Hoyningen-Huene would recommend), but with a more robust notion of phenomena than mere visual gestalt switches (as Bird enjoins), and yet still perspectival in nature (as Giere would suggest). Am I trying to square the circle? Maybe. But let us see first what NKKs are, before we can evaluate their promise.

Two intuitions underpin NKKs (for details, please see Massimi 2014a). The first comes from Quine, and the second from Kant. Recall Quine's (1968) indeterminacy of reference and ontological relativity resulting from a plurality of manuals of translation. If there is no fact of the matter about the reference of the native's term "gavagai" and there could be a plurality of manuals of translations (or, as Kuhn would say, a plurality of *interpretations*), what is there to be said about the kind "rabbit"? What about emeralds, ravens, water, or electromagnetic field? Yet we do make inductive inferences about those objects; we reach universal generalisations about them; we can even make scientific predictions about them. We need a notion of natural kind that can serve the purpose of induction, prediction, and explanation in science. Quine famously offered one, Darwinian in spirit, in claiming that our successful inductive inferences are the result of trial-and-error and the kinds that survive are those, which prove congenial to successful induction. Our subjective spacing of qualities proves survival-adaptive and accords so well with what Quine called "functionally relevant groupings in nature" to make our inductive inferences come out right (Quine 1969, p 126). Yet, our inductive inferences come out right not because our subjective spacing of qualities tracks well-defined kinds in nature. Instead, by trial-and-error new groupings prove favorable to induction and become "entrenched". In Quine's words, "in induction, nothing succeeds like success" (Quine 1969, p. 129). Thus, this is the first Quinean intuition about NKKs: our kinds are projectable and prove favorable to induction, not because they carve nature's joints, but rather because they consist in "functionally relevant groupings" that prove survival-adaptive, and thus become entrenched.

The second intuition behind NKKs is Kantian in claiming that what does the 'grouping' is neither nominalist convention nor social construction. Instead, the 'grouping' reflects epistemic constraints that *we* impose on nature qua cognitive agents. That we cluster some empirical properties (rather than others), give them names, and designate *kinds* of objects with them, is neither a matter of arbitrary labeling nor of mob psychology. Our kinds are the expression of our conditions of possibility of knowledge, of what is possible for cognitive agents like us to *know* about nature. Kant famously used transcendental arguments to elicit the categories that he thought may provide the conditions of possibility of knowledge for us. Transcendental arguments take the following form (for a discussion, see Stern 1999):

a. We have experience of X.
b. For X to be an object of experience for us, it must exhibit feature A.
c. Therefore, X exhibits A.

For example, X could be inelastic collisions between two hard bodies (say, two billiard balls). Premise b. tells us that for us to be able to experience this phenomenon, the phenomenon must exhibit some feature A. For example, A could be causality, or the idea that *causes equal their effects*—i.e. the overall momentum of the incident ball should equal the overall momentum of the outgoing ball (where, despite kinetic energy being transformed and dissipated in inelastic collisions, we must assume momentum to be conserved so we can calculate the final velocities of the balls). Therefore, we conclude that this is the case (as per premise c.). On the basis of transcendental arguments, Kant arrived at his table of categories of the understanding (whereby causality for Kant was one of the transcendental principles in the category of relation).

Thus, the second Kantian intuition behind NKKs simply tells us that for us to have knowledge of natural kinds (as we do have it), certain epistemic conditions must be met. For example, we might have to think of the grouping in question (e.g. inelastic collisions) as being governed by a causal relation (i.e. conservation of momentum), on pain of Humean skepticism about causation and induction. Causation is one among other examples of conditions of possibility of experience, including also action and reaction (e.g. mechanical motions subject to Newton's third law) or having a magnitude that comes in degrees. Obviously, one does not need to hark back to Kant's outmoded categories of the understanding to make the point I want to make here. The intuition is sufficiently simple: for us to have knowledge of the gravitational field, inelastic collisions, mechanical motions, and so on, as *natural kinds*, we must assume that they have intensive magnitudes, enter into cause-effect relations, obey action and reaction laws, and so forth. But these features are neither essential features (qua dispositional powers of the objects in nature) nor sheer nominal labels. They are not constructions of laboratory life either, as constructivists would maintain. They are instead *epistemic conditions* that Quinean groupings of empirical properties, which we call "gravitational field", "inelastic collisions", and so on, have to meet for them to become objects of scientific knowledge.

The two (Quinean and Kantian) intuitions jointly give us a definition of naturalized Kantian kinds:

> (NKK): Scientific kinds are groupings or clusters of empirical properties, which have proved survival-adaptive *and* have met our conditions of possibility of experience (but not via some constructive activity of our mind).

Interestingly, while the Quinean condition accounts for the resilience and inductive success of our scientific kinds, the Kantian condition explains why we come to know certain kinds but not others (which might also be compatible with the same clusters of empirical properties). Moreover, it explains why our kinds *do* change over time, when our conditions of possibility of experience change (after a scientific revolution), following the Kant-on-wheels route mentioned above. I have discussed the prospects of NKKs in relation to Boyd's criticism of neo-Kantian accounts of natural kinds elsewhere (Massimi, 2014a). In the remaining pages of this paper, I'd like to illustrate the advantages of endorsing NKKs to understand Kuhn's world-changes, and to address the aforementioned challenges faced by both Hoyningen-Huene's and Giere's account.

10.5   NKKs to the Rescue

In Sect. 10.2, we discussed Hoyningen-Huene's Kantian reading of Kuhn. If we understand Kuhn's controversial claim about world-changes as a claim about phenomenal worlds (rather than noumena), the position no longer seems as paradoxical as it would otherwise seem. Yet, as Bird notes, the Kantian reading of Kuhn is difficult to square with Kuhn's interest in psychology and cognitive science, *and* Kantian phenomena cannot be identified with visual gestalt switches. I called these the *challenge of naturalism* (CoN) and the *challenge of phenomena* (CoP). Equipped with the above definition of NKKs, we can now return to these challenges and see how they both dissolve.

As is should be clear from the discussion in the previous Section, the *challenge of naturalism* (CoN) does not arise at all. Recall that CoN arose under the Kantian reading because if we identify the physical world with a noumenal world of things-in-themselves we cannot explain how the phenomenal world (with its scientific kinds) can possibly be the product of our causal interaction with the physical world (via sense data, stimuli, background beliefs and so forth). The problem does not arise at all if we interpret scientific kinds as NKKs. For, in this case, naturalism has been built into the definition of NKKs from the ground up. Our scientific kinds are groupings or clusters of empirical properties that we have identified in nature as "functionally relevant", to borrow Quine's apt expression. No mysterious interaction between the phenomenal world and things-in-themselves is involved in NKKs.

Similarly, the *challenge of phenomena* (CoP) dissolves too. The problem in that case was that *phenomenal worlds* are not reducible to psychological gestalt switches and that a proper Kantian reading would require a more robust notion of phenomena than visual or sensory experiences. NKKs can take care of this feature. NKKs (be it inelastic collisions, gravitational field, or other) count as Kantian phenomena that have met *conditions of possibility of experience,* as explained in the previous section (i.e., we must think of the gravitational potential as having intensive magnitude; as much as we must think of inelastic collisions as satisfying conservation of momentum as a principle of causality; and, so forth). Thus, NKKs are more than mere visual gestalt switches, as the Kantian reading would have it, pace CoP.

Not only can NKKs handle the challenges faced by the Kantian reading of Kuhn. They can also deliver on the promise of a perspectival realist reading of Kuhn, without having to pay the hefty price of either *conceptual relativism* (CR) or *alethic relativism* (AR), as outlined in Sect. 10.3. Following up on Giere's two versions of perspectivism in Kuhn, we can now see how NKKs are compatible with either version (PiKu$_1$ and

PiKu$_2$) while also avoiding the problem that each version faces. NKKs can be taken as the products of taxonomic categories, defining scientific perspectives, along the lines of PiKu$_1$. We can think of, say, inelastic collisions as the product of taxonomic categories of eighteenth-century Leibnizian dynamics, (e.g., where the Leibnizian principle of *causes equal their effects* found its expression in the conservation of *vis viva*). Inelastic collisions constitute then a *kind* of mechanical motion, characterized by the scientific perspective of eighteenth-century Leibnizian dynamics, for example. No wonder that scientists studying inelastic collisions these days 'live and work in a new world'—the taxonomic categories available to Leibniz and contemporaries (e.g. from elasticity to living force) are no longer available to us for interpreting these empirical regularities. Thus, inelastic collisions are NKKs in so far as they are empirical regularities in nature that have been grouped or clustered according to some epistemic principle (such as, for example, causality) that has proved resilient despite major changes from Leibnizian dynamics to contemporary physics.

While a NKK take on collisions is compatible with the PiKu$_1$ reading, it has also the advantage of avoiding the *conceptual relativism* (CR) that seems to affect the latter. Recall that the pitfall of PiKu$_1$ was that scientific perspectives were identified with conceptual taxonomies cutting boundaries into the worldly dough to the extent that, beyond the conceptual taxonomy of a scientific perspective, there does not seem to be any well-defined fact of the matter. Thus, one can interpret Kuhn's world-changes along PiKu$_1$ lines as saying that when we switch from Leibnizian dynamics to relativistic mechanics, the very *kind* of motion that we call "inelastic collisions" changes, because of a dramatic change in the taxonomic categories and their associated laws. Beyond these two scientific perspectives with their respective conceptual taxonomies, there is no fact of the matter about inelastic collisions as a *kind* of motion (although there could well be a fact of the matter about individual balls moving, along a nominalist view that Kuhn never regarded as his own). This is the CR that seems to ensue from PiKu$_1$. World-changes are ultimately *kind-changes* induced by changes in the conceptual taxonomies of scientific perspectives.

NKKs seem exempt from the threat of conceptual relativism. The threat is a genuine one for PiKu$_1$ because conceptual taxonomies are taken to be cookie-cutters in the worldly dough. But if instead of entrusting conceptual taxonomies with this *metaphysical* role of drawing boundaries for scientific kinds, we entrust them only with the *epistemic role* of providing conditions of possibility of experience (as per NKKs), CR is averted. I have discussed and criticized elsewhere (Massimi 2014b) the seemingly metaphysical role that conceptual categories seem to play in Kuhn and the resulting constructivist readings of Kuhn that have occasionally been offered. Here, I want to reiterate that such constructivist readings can be avoided, and must be avoided. NKKs provide an alternative way of thinking about world-changes along non-constructivist lines. World-changes qua *kind-changes* do still depend on us and on our scientific perspectives, but not in virtue of any constructive activity of the conceptual taxonomy at issue.

NKKs are also compatible with the PiKu$_2$ reading. Our scientific kinds can be regarded as the products of theoretical principles constitutive of a lexicon, via which we gain *experience of nature*. We could take, for example, the gravitational field as a NKK, which in classical mechanics does not denote any genuine field (as gravity is supposed to be a force acting at a distance as given by Newton's law of gravity); whereas, in general relativity, it denotes a genuine field whose behavior is described by Einstein's field equations. Under PiKu$_2$, we can take Newton's law of gravity and Einstein's field equations as *constitutive* of two alternative scientific lexicons or perspectives. Scientists 'live in a new world' after general relativity because the same *kind* of field (e.g. the gravitational field) is no longer thought of as a fictional vector field surrounding a point-like body of mass m (to which Newton's law applies). It is instead thought of as a curved space-time, determined by the distribution of matter and energy (as given by the stress-energy tensor). The gravitational field is a NKK insofar as it consists of empirical regularities (associated with the gravitational behavior of bodies in nature) clustered according to theoretical principles (be it Newton's law of gravity or Einstein's field equations), which are *constitutive of the experience* within their respective scientific perspectives.

NKKs are then compatible with the second perspectival reading (PiKu$_2$) of Kuhn, while also avoiding the charge of alethic relativism (AR) that affects it. Recall that AR arises if we take constitutive theoretical principles to define what we can *truly assert* about the world. Thus, scientific lexicons/ perspectives can be regarded as defining the truth or falsity of propositions. Truth claims (about, say, the gravitational field) become relative to perspectives. That gravity is a force acting at a distance among point-like masses is true in Newtonian mechanics, but false in general relativity. This much may be granted. But relativism about truth is at odds with realism. If Kuhn's position stands a chance of being called realist (or perspectival realist), I argue that truth cannot be indexed to scientific perspectives: there cannot be facts about the very same object or kinds of objects, which are true in one perspective and false in another one (for a defense of indexing truth to perspectives, see Hales 2006). Perhaps Kuhn himself was willing to accept that much. Perhaps his position has been presented and reconstructed as implying that much. World-changes seem to entail *truth-changes* induced by a change in the theoretical principles constitutive of scientific perspectives.

But I suggest that we should resist this conclusion, and NKKs provide an antidote against AR. The truths that stay the same across scientific perspectives are the empirical regularities and properties that NKKs track in nature. Following Quine's Darwinian intuition, these are the regularities that have proved inductively successful, and survival adaptive. They have become entrenched into the scientific kinds we know and love. And this much we can truly assert, *independently of* which scientific perspective we happen to work within. The theoretical principles constitutive of a perspective define what we can reasonably *come to know* about those clusters of empirical properties and regularities. They play an *epistemic role* in defining the conditions of possibility of our experience: what we may be *justified to believe* about those clusters of properties. But they do not define nor determine what is true of those empirical properties. *That* there are those empirical properties, *that* they do track empirical regularities in nature may or may not have an ultimate explanation. But that some clusters prove favorable to induction and become entrenched, while others systematically wrong have a tendency to get discarded, is a fact about nature and its 'functionally relevant grouping', which is invariant across scientific perspectives. These are the truths that resist the challenge of AR. World-changes qua *kind-changes* do still depend on us and on our scientific perspectives, without any need of indexing truth and falsity (or truth-makers) to scientific perspectives.

10.6 Conclusion

I hope I have achieved two goals in this paper. The first was to show that a Kantian reading of Kuhn chimes with a more recent perspectival realist reading offered by Giere. If we take scientific perspectives to roughly correspond to Kuhn's scientific paradigms/lexicons, two perspectival readings of Kuhn become available, both of which take world-changes as changes in phenomenal worlds (as Hoyningen-Huene's Kantian reading would have it). Under the first perspectival reading of Kuhn (which I called $PiKu_1$), world-changes are changes in the conceptual taxonomies of a scientific perspective. Under the second perspectival reading ($PiKu_2$), world-changes are changes in the theoretical principles constitutive of a scientific perspective.

But, both the Kantian reading and the perspectival realist readings face some problems. My second goal in this paper was to illustrate these problems and to propose a solution in terms of naturalized Kantian kinds (NKKs). I explained the two (Quinean and Kantian) intuitions behind NKKs, and showed how they can come to the rescue with some of the problems affecting the Kantian and perspectival readings of Kuhn. World-changes become *kind-changes* where scientific kinds consist of clusters of empirical properties grouped by scientific perspectives with their resources, which we take to be *constitutive of our experience* of nature. More work has to be done to clarify exactly the nature of NKKs, the epistemic role played by conceptual taxonomies and theoretical principles, and how their historical contingency does not open the door to unwelcome forms of relativism (either conceptual or alethic). But I hope the present paper offers a first step towards exploring these pressing and still open questions, fifty years after Kuhn's trailblazing book.

Acknowledgments I thank Ron Giere, Paul Hoyningen-Huene and Paul Teller for kindly reading earlier versions of this paper. Not surprisingly perhaps, they did not share my worry about relativism and incommensurability; but they were all very gracious to provide me with helpful comments and suggestions for improving the paper.

References

Bird, A. 2000. *Thomas Kuhn*. Princeton: Princeton University Press.
Bird, A. 2012. *The structure of scientific revolutions* and its significance: An essay review of the fiftieth anniversary edition. *British Journal for the Philosophy of Science* 63:859–883.
Chignell, A. (2010) Real repugnance and belief about things-in-themselves. In *Kant's moral metaphysics,* ed. J. Krueger and B. Bruxvoort Lipscomb. Walter DeGruyter.

Friedman, M. 2001. *The dynamics of reason*. CSLI Publications.

Giere, R. 2006. *Scientific perspectivism*. Chicago: University of Chicago Press.

Giere, R. 2013. Kuhn as perspectival realist. *Topoi* 32:53–57.

Hacking, I. 1993. Working in a new world: The taxonomic solution. In *World changes: Thomas Kuhn and the nature of science,* ed. P. Horwich, 275–310. Cambridge: MIT Press.

Hales, S. 2006. *Relativism and the foundations of philosophy*. Cambridge: MIT Press.

Hanson, N. R. 1958. *Patterns of discovery*. New York: Cambridge University Press.

Hoyningen-Huene, P. 1993. *Reconstructing scientific revolutions*. Chicago: University of Chicago Press.

Hoyningen-Huene, P. 2008. Commentary on Bird's paper. In *Rethinking scientific change and theory comparison,* ed. L. Soler, H. Sankey and P. Hoyingen-Huene, 41–46. Dordrecht: Springer

Kuhn, T. S. 1962. *The structure of scientific revolutions*. Chicago: University of Chicago Press.

Kuhn, T. S. 1977. *The essential tension*. Chicago: The University of Chicago Press.

Kuhn, T. S. 1990. Dubbing and re-dubbing: The vulnerability of rigid designation. In *Scientific theories,* Studies in the Philosophy of Science XIV ed. C. Wade Savage, 298–318. Minneapolis: University of Minnesota Press.

Kuhn, T. S. 1991. The road since structure. In *PSA 1990 proceedings of the 1990 biennial meeting of the Philosophy of Science Association,* vol. 2, ed. A. Fine, M. Forbes, and L. Wessels, 90–105. East Lansing: Philosophy of Science Association. (Reprinted in Kuhn (2000)).

Kuhn, T. S. 1993. Afterwords. In *World changes: Thomas Kuhn and the nature of science,* ed. P. Horwich, 224–253. Cambridge: MIT Press, 311-39. Reprinted in Kuhn (2000).

Kuhn, T. S. 2000. *The road since structure. Philosophical essays, 1970–1993, with an autobiographical interview*. Chicago: University of Chicago Press.

Langton, R. 1998. *Kantian humility*. Oxford: Clarendon Press.

LaPorte, J. 2004. *Natural kinds and conceptual change*. Cambridge: Cambridge University Press.

Lipton, P. 2001. "Kant on wheels", Review of *The road since structure. London Review of Books* 23 (14): 30–31.

Massimi, M. 2014a. Natural kinds and naturalised Kantianism. *Nous* 48 (3): 416–449.

Massimi, M. 2014b. Working in a new world: Kuhn, constructivism, and mind-dependence. *Studies in History and Philosophy of Science*, doi: 10.1016/j.shpsa.2014.09.011.

Quine, W. V. O. 1968. Ontological relativity. *Journal of Philosophy* 65:185–212.

Quine, W. V. O. 1969. Natural kinds. In *Ontological relativity and other essays,* ed. W. V. O. Quine, 114–138. New York: Columbia University Press.

Stern, R. 1999. *Transcendental arguments: Problems and prospects*. Oxford: Clarendon Press.

Teller, P. 2011. Two models of truth, *Analysis* 71, 465–472.

Chapter 11
An Analysis of Truth in Kuhn's Philosophical Enterprise

William J. Devlin

11.1 Introduction

In 2012, Thomas Kuhn's *Structure of Scientific Revolutions* (henceforth, *Structure*) celebrated its 50th anniversary. Kuhn's work was groundbreaking in many respects— from the introduction of the notions of a *paradigm-shift* to *incommensurability*, to a novel picture of scientific development, *Structure* became influential, at the very least, in the fields of philosophy, science, history, sociology, psychology. One of the enduring legacies of Kuhn's *Structure* 50 years on is his rejection of the view that science is progressing towards the truth.

Kuhn challenges this view by focusing particularly on the traditional account of the correspondence theory of truth. As he explains in his 1969 Postscript to *Structure*, "[o]ne often hears that successive scientific theories grow ever closer to, or approximate" more and more closely to, the truth ... [in the sense that scientific theories depict a] match, that is, between the entities with which the theory populates nature and what is really there." However, as Kuhn continues to suggest, the traditional correspondence theory of truth fails on the grounds that we cannot epistemically access 'what is really there': "There is, I think, no theory-independent way to reconstruct phrases like 'really there'; the notion of a match between the ontology and its "real" counterpart in nature now seems to me illusive in principle." As such, scientific development should not be understood as a linear progression towards the truth about the mind-independent world. Instead, scientific development is best construed as following a cyclical pattern of paradigm, normal science, crisis, revolution, paradigm-shift. In this framework, Kuhn fundamentally rejects the traditional correspondence theory of truth, and suggests that truth no longer plays a pivotal role in science: "Perhaps there is some other way of salvaging the notion of 'truth' for application to whole theories, but this one will not do." (Kuhn 1962/1970, p. 206).

W. J. Devlin (✉)
Department of Philosophy, Bridgewater State University,
Tillinghast Hall, 02325 Bridgewater, MA, USA
e-mail: wdevlin@bridgew.edu

© Springer International Publishing Switzerland 2015
W. J. Devlin, A. Bokulich (eds.), *Kuhn's Structure of Scientific Revolutions—50 Years On*,
Boston Studies in the Philosophy and History of Science 311,
DOI 10.1007/978-3-319-13383-6_11

Decades after *Structure*, Kuhn continued to focus his philosophical attention on criticizing the notion of truth in science—so much, that he even considered it to be one of the central goals of his philosophy of science, or "the nature of his philosophical enterprise":

> My goal is double. On the one hand, I aim to justify that science is cognitive, that its product is knowledge of nature, and that the criteria it uses in evaluating beliefs are in that sense epistemic. But on the other, I aim to deny all meaning to claims that successive scientific beliefs become more and more probable or better and better approximations to the truth and simultaneously to suggest that the subject of truth claims cannot be a relation between beliefs and a putatively mind-independent or 'external world'. (Kuhn 1993, pp. 329–330)

We can divide Kuhn's double-goal into three distinct claims. Kuhn's first goal (call this G_1) can be formulated as follows: 'To show that science is cognitive; it achieves knowledge of nature; the criteria it uses are epistemic'. Meanwhile, Kuhn's second goal (call this G_2) can be divided into sub-goals. On the one hand, Kuhn aims to show that 'Scientific beliefs are not directed towards truth (exact or approximate)' (call this sub-goal, G_{2a}). On the other hand, Kuhn aims to show that 'the subject of truth claims is not a relation between belief and the independent world' (call this sub-goal, G_{2b}).

Kuhn laid the groundwork for accomplishing this "double-goal" in *Structure*, which—through the introduction of his account of paradigms, scientific revolutions, and incommensurability—helped to shape new horizons in the philosophy of science. *Structure* has traditionally been understood as providing a justification of Kuhn's theses via the 'pessimistic meta-induction' argument from past falsity' and the 'theory-ladenness of observation', where Kuhn examines several 'key examples' in the history of science to show that scientific theories are not progressing towards truth; rather, science operates through paradigms, anomalies, crises, revolutions, and paradigm shifts in which there is no coherent ontological development. For Kuhn then, science does not work towards a notion of truth. Likewise, we do not know if science has ever been able to successfully make claims about the world as it really is, independent of human consciousness (i.e., the 'independent world'). Because we are unable to verify such claims, Kuhn maintains that science should abandon the attempt to make claims about the independent world.

However, as I will show, Kuhn's denial of truth helps to bring out a significant problem between the two goals of his enterprise. More specifically, it appears as though Kuhn cannot both maintain that science achieves knowledge of nature (G_1) and dismiss the notion of truth altogether from his philosophy of science (G_{2a}). The same arguments that aim for the dismissal of truth from scientific inquiry will ultimately challenge the quest for scientific knowledge.

In this chapter, I explore this difficulty—which I will call the *problem of inconsistency*—and ultimately provide a defense on behalf of Kuhn. I argue that Kuhn can achieve both goals of his 'double goal', and so remain consistent, by re-evaluating his rejection of truth in his philosophy of science. This defense will be driven by introducing an alternative version of the correspondence theory of truth that differs from the 'traditional' correspondence theory (CTT) that Kuhn rejected. I will argue that, even though Kuhn rejects CTT, his philosophy of science will remain consistent with an alternative notion of correspondence. I believe that this notion of truth not only succeeds in defending Kuhn's enterprise from the problem of inconsistency, but is also a kind of truth that Kuhn can accept into his philosophical enterprise.

In order to see how this alternative theory succeeds in defending Kuhn from the problem, I will first review CTT and Kuhn's rejection of the theory. Next, I will turn to the problem of inconsistency that arises due to Kuhn's dismissal of truth. Finally, I will present the alternative view of correspondence and examine how it both fits comfortably into Kuhn's enterprise and overcomes the problem of inconsistency.

11.2   The Correspondence Theory of Truth

Kuhn's attempt to achieve the second goal (G_2) of his double goal— that 'scientific beliefs are not directed towards exact or approximate truth' (G_{2a}) and 'the subject of truth claims is not a relation between belief and the independent world' (G_{2b})— centers on his analysis of CTT. Kuhn summarizes CTT in the following two passages:

> Within the main formulation of the previous tradition in philosophy of science, beliefs were to be evaluated for their truth or for their probability of being true, where truth meant something like corresponding to the real, the mind-independent external world (1991, p. 114)
> [T]he tradition supposed that good reasons for belief could be supplied only by neutral observations, the sort of observations, that is, which are the same for all observers and also independent of all other beliefs and theories. These provided the stable Archimedean platform required to determine the truth or the probability of the particular belief, law, or theory to be evaluated. (1991, p. 113)

From these two passages, we can extrapolate a specific formulation of CTT. This formulation essentially entails at least two metaphysical claims, which can be formulated as follows:

- P_1: Truth is a property of statements.
- P_2: A statement is true if and only if it corresponds to a fact that obtains in a world *independent of our cognitive awareness (i.e., the independent world).*

P_1 seems to be, for the most part, straightforward, and is a standard characteristic of CTT, so let's leave it be. P_2, however needs to be clarified. More generally, we could formulate P_2 as: "A statement or proposition is true if and only if it corresponds to a fact that obtains." But if so, we need to ask ourselves, 'What does it mean to be a *fact that obtains?*' One interpretation is that a fact is something that is contained in our world of experience; i.e., the empirical world containing sense data—a world dependent upon our cognitive mental processes. However, most supporters of CTT wish to extend this notion of facts that obtain to the mind-independent world, or the world independent of our cognitive mental processes. For this reason, then, P_2 should specify that: 'A proposition is true if and only if it corresponds to a fact that obtains in a world *independent of our cognitive awareness.*'

Together, P_1 and P_2 make up the central claims of CTT and can be considered the metaphysical claims of the theory. But those who espouse CTT are committed to at least one further claim. In order to make this theory applicable, we must establish a criterion for the distinction between knowledge and false beliefs. We can formulate the corollary claim to CTT, call this C_1, as: 'We have epistemic access to the mind-independent world which allows us to determine the truth-value of our statements'. C_1, however, is not enough to make CTT applicable. Indeed, epistemologically, we do need accessibility to the independent world in order to distinguish between knowledge and beliefs. But having epistemic accessibility is not enough—we still need a medium that can represent both our beliefs and the world so that they can be compared. Such a medium is ready-made in language. And, given that our theory concerns correspondence, it follows that we would like language to be able to represent the facts that obtain in the independent world. So, the second corollary claim to CTT, call this C_2 can be formulated as: 'Language has the capacity to represent the facts that obtain in the mind-independent world'.

In sum, CTT (or at least the traditional formulation relevant to Kuhn's criticism of CTT) contains at least two central metaphysical claims, and two corollary claims:

- **P_1**: Truth is a property of statements.
- **P_2**: A statement is true if and only if it corresponds to a fact that obtains in a world *independent of our cognitive awareness (i.e., the independent world)*.
- **C_1**: We have access to the independent world insofar as our world of experience is caused by and faithfully represents the independent world.
- **C_2**: Language has the capacity to represent the facts that obtain in the independent world

11.3   Kuhn's Road to Separating Science From Truth

As we saw in Sect. 11.1, Kuhn's second goal to separate science from truth can best be understood as unfolding in two parts or sub-goals: first, to show that 'Scientific beliefs are not directed towards exact or approximate truth' (G_{2a}); and second, to show that 'the subject of truth claims is not a relation between belief and the independent world' (G_{2b}). In order to set up the problem of inconsistency and the resolution I wish to argue for, it is helpful to review the roads, or arguments, Kuhn follows to achieve both G_{2a} and G_{2b}. I will begin with the latter road, followed by the former.

11.3.1   Kuhn's Rejection of CTT

Kuhn's justification of G_{2b} concerns his rejection of CTT. His central reason for rejecting CTT is his dismissal of C_1, the epistemic claim of the traditional correspondence theory. We find that Kuhn sets up C_1 as follows:

> [T]he tradition supposed that good reasons for belief could be supplied only by neutral observations, the sort of observations, that is, which are the same for all observers and

also independent of all other beliefs and theories. These provided the stable Archimedean
platform required to determine the truth or the probability of the particular belief, law, or
theory to be evaluated. (1991, p. 113)

In this passage, Kuhn presents the epistemic claim of CTT as involving neutral observations, or observations that are independent from the influences of both theory
and belief, that help to provide the independent access—the stable "Archimedean
platform"— to judge whether or not our statements and judgments (for Kuhn, particular beliefs, laws, or theories) correspond to facts that obtain in the independent
world. Thus, we can see that Kuhn presents three of the four claims that make up
CTT (Kuhn 1991, p. 95, 1991, pp. 113–116).

Kuhn proceeds to attack the tenability of C_1. His attack focuses on the claim that
we have access to a stable Archimedean platform—an epistemic vantage point that
has traditionally assumed to be accessible. According to Kuhn, we do not have a
justified procedure for connecting our beliefs and theories to the independent world:
"Seldom or never can one compare a newly proposed law or theory directly with
reality (Kuhn 1991, p. 114). The problem is that we cannot compare our theories
or beliefs about the world with 'reality' because "Only a fixed, rigid Archimedean
platform could supply a base from which to measure the distance between current
belief and true belief." For Kuhn, then, the problem with C_1, is that it is impossible
to stand outside of our own theory and beliefs, or point of view, and see the world as
it is, independent of our cognitive awareness. Kuhn maintains that we lack the fixed
Archimedean platform upon which to stand, and so we cannot connect our world
of experience to the independent world: ". . . the Archimedean platform outside of
history, outside of time and space, is gone beyond recall" (1991, p. 115). Without
this platform, Kuhn maintains that we do not have a neutral place in which to stand
to evaluate whether or not our statements and propositions correspond to facts that
obtain in the independent world.

This critique echoes Hilary Putnam's attack upon the CTT where Putnam argues
that human beings are not in a position to judge whether or not our statements
correspond to the independent world: "To single out a correspondence between two
domains, one needs some independent access to both domains" (Putnam 1981, p.
74). The problem for Putnam is that we do not have this independent access—we do
not have a place to stand (a 'God's eye view') to judge whether or not our statements
correspond to facts in the independent world. Kuhn makes the same point: in order to
say that our statements are true for the independent world, we must be in a position
to determine whether or not our statements are true for the world; we must be able to
stand outside of our perspective and view the world of experience and the independent
world from a 'perspectiveless perspective'.[1] Such a stance, however, is not possible
for Kuhn—we cannot elevate ourselves above the world we experience [which is
partially determined by paradigms] and determine whether or not our statements
correspond to an independent world.

---

[1] For a discussion of Kuhn in relation to perspectival realism, see Massimi, Chap. 10, this volume.

In other words, Kuhn argues that if we are to have access to the facts that obtain in the independent world, then it must be through the neutral ground of an a-historical, a-temporal, and a-spatial Archimedean platform. Such a platform is the *only* way to come to know whether or not our statements correspond—this epistemic position is a necessary condition for accessibility to facts that obtain in the independent world, according to Kuhn. Now, given that Kuhn rejects the existence of any platform, it follows that we do not have epistemic accessibility to the independent world. Thus, Kuhn dismisses CTT on the grounds that we cannot accept C_1.

11.3.2 Kuhn's Science Without Truth

Kuhn's justification of G_{2a} ('scientific beliefs are not directed towards exact or approximate truth') is rooted in *Structure*, where he develops several essential concepts that characterize the pattern of scientific change, including paradigm, normal science, crisis, scientific revolution, and incommensurability. Kuhn claims that science operates under *paradigms*, which Kuhn breaks up into two separate concepts. The first notion, the *paradigm as disciplinary matrix*, is a broader, more globalized notion of paradigm insofar as it consists of interrelated conceptual, theoretical, instrumental, and methodological commitments within scientific practice that include symbolic generalizations, metaphysical beliefs, epistemic values and *shared exemplars* (Kuhn 1962/1970, pp. 182–186), which become the second, and for Kuhn more fundamental, notion of paradigm (the *paradigm as shared exemplar*). The latter kind of paradigm is a specific, or localized paradigm, in that it serves as the example for how to solve scientific problems of the global paradigm. The shared exemplar is crucial to the normal activities of science, or *normal science*, which is understood as 'everyday science', where the goal is to mop-up and "puzzle solve" those lingering questions and thereby extend the body of knowledge of the facts about nature, given the disciplinary matrix. For Kuhn, normal science is analogous to 'puzzle solving'. As someone approaching a jigsaw puzzle, the scientist begins with an already established framework—one is already given a set of pieces to work with, each of which has a given place within that framework. The scientist follows the guidelines to make the pieces fit. There is a solution and the scientist is guaranteed an answer to the question that she is asking. Kuhn acknowledges that this answer is already partially determined by the framework provided by the disciplinary matrix, and so the framework tells us what will count as a possible solution to the problem that normal science investigates. However, as Kuhn argues in *Structure,* normal science continues until a *crisis*, or a decadent state of the elements in the paradigm caused by an unresolved anomaly, which leads to the possibility of a paradigm-shift, or a *scientific revolution*, whereby, through a new shared exemplar, a new and incompatible paradigm replaces the previous one.

Kuhn clarifies the "incompatibility" between the older and new paradigm through the introduction of the thesis of *incommensurability*, which holds that "proponents of competing paradigms must fail to make complete contact with each other's view-points" as they "are always at least slightly at cross-purposes." Cross-purposes between paradigms occur for three reasons. Following Howard Sankey, (1993) we can label these reasons as (i) standard variance, (ii) conceptual disparity, and (iii) the theory-dependence of observation. First, paradigms are different in terms of standard variance where "[t]he proponents of competing paradigms ... often disagree about the list of problems that any candidate for paradigm must resolve. Their standards or their definitions of science are not the same" (Kuhn 1962/1970, p. 148). For instance, Kuhn notes that proponents of Aristotelian science consider that if science includes a theory of motion, it must explain the cause of the attractive forces between particles of matter. Meanwhile, Newton's dynamics implies a notion of science where it is not necessary to provide such an explanation; rather Newton's conception of science considers it sufficient to only note the existence of such forces. The proponents of the competing 'Aristotelian' and 'Newtonian' paradigms thus disagree about whether or not the explanation of the cause of attractive forces is a problem that must be resolved. In this sense, proponents of competing paradigms cannot find a common neutral 'ground' upon which to stand to determine which list of problems are correct to serve as a standard of science. Second, paradigms differ in terms of conceptual disparity insofar as each paradigm has a different network of relations between concepts, resulting in different ways for how the concepts are used, or attached to nature. Without recourse to a neutral position outside of a paradigm, there cannot be a direct and complete translation of concepts (and concept use) without residues or loss of meaning of the original terms to be translated. Third, paradigms differ as a result of the theory-dependence of observation, which Kuhn explains as occurring insofar as "the proponents of competing paradigms practice their trades in different worlds." The theoretical and conceptual components of a given paradigm affect the paradigm-members' processes of observation, so that members of different paradigms experience worlds radically different from one another—they see different things and conceptually organize their sense-data differently.

The thesis of incommensurability suggests that there is not only a gap between paradigms, but also a gap between any given paradigm and the independent world. For Kuhn, the paradigm shapes the scientist's view of the world—it determines not only how the scientist observes and experiences the world, but also how the scientist knows the world. The scientist's world is a world that is thus structured by, and partially dependent upon, the conceptual scheme of the paradigm. Here, we can see Kuhn's application of his criticism of C_1, into science. The scientist does not reach the independent world insofar as he does not have access to the fixed Archimedean platform that is necessary to gain epistemic access to the independent world. Even though the normal scientist believes she is making judgment claims about the facts that obtain in the independent world, she is epistemologically limited by her paradigm. Given this limitation, the scientist is not in a position to judge which statements and theories are true.

While Kuhn rejects C_1, he does not leave us without any epistemic basis for evaluating scientific knowledge claims. He offers us an 'epistemic substitute' for the fixed Archimedean platform, which he calls the "moving Archimedean platform"

(Kuhn 1990, p. 95, 1991, p. 113). The only ground we can stand upon for knowledge of scientific claims is a platform that moves: it does not transcend paradigms to serve as the universal platform that is required to achieve knowledge of the independent world. Instead, the platform is moving insofar as there is a different platform for each paradigm—what counts for success of science and scientific knowledge is limited by the historical paradigm in which science belongs. In this way, Kuhn drops the scientific drive towards truth, and replaces it with the more localized goal of puzzle solving. The justification is internal to the paradigm insofar as the Archimedean platform 'moves' along with the paradigm shift.

11.4 The Problem of Inconsistency

Thus far, we have followed Kuhn's roads to achieve the second goal of his philo-sophical enterprise (G_{2a} and G_{2b}), both of which involve his criticism of CTT. But now that we have a clearer picture of these roads, I suggest that Kuhn has trouble achieving both of his sub-goals, as he is tangled up between the two. He seems to think that his epistemic critique of the CTT shows that he has defeated this account of truth (G_{2b}) and then, given the loss of CTT, we can remove truth entirely from science (G_{2a}). If this is Kuhn's view, then he is making two mistakes. First, he is wrong to hold that his epistemic critique shows that truth is not a relation between statements and the independent world. Kuhn's epistemic argument against CTT does not show us that the metaphysical claims about truth (P_1 and P_2) are incorrect; rather, Kuhn has shown us that the *utility* of truth as promoted in CTT is in danger, since we do not have epistemic access. Thus Kuhn's epistemic critique does not show us that truth is not a relation between beliefs and the independent world. At best, he has shown us that humans are unable to know the truth about the world, where truth may still be under CTT.

Second, Kuhn is mistaken to believe that his epistemic critique, even if successful, shows that we should remove truth from science. This mistake holds especially if Kuhn thinks that an argument showing that the CTT is incorrect ultimately leads to the dismissal of truth from science. One can easily respond that if we show that truth incorrectly depicted under CTT, then we are still not ready to remove truth from science until we know exactly what truth is. At best, Kuhn has shown that the kind of truth discussed in CTT—what we can call 'absolute truth' where truth is universal and timeless, as it belongs to a relation to the (fixed and static) independent world, is useless (given our lack of accessibility to the independent world), and so should not be part of science. But before Kuhn can dismiss truth entirely from science, he must be sure that none of the other theories of truth provide a more accurate depiction of what truth is, and if they do, that they too are useless.

Thus, it appears that while Kuhn seems to create a growing tension between the two sub-claims of his second goal, he does at least suggest that the utility of the kind of absolute truth as it is in CTT is questionable, and if that is truth, then it may be best to see that truth as such should be removed from science. Here, we can see that

Kuhn is at least able to begin his pursuit for achieving the first sub-claim (G_{2a}) of his second goal.

Still, I argue that Kuhn faces a larger problem concerning his goal to dismiss truth entirely from the project of science. Notice that one of the claims of G_1 is that science achieves "knowledge of nature." Given Kuhn's position towards rejecting the role of truth in science, we have a tension between G_1 and G_2: Kuhn wants to say that science doesn't achieve the truth about nature, yet at the same time, science *does* achieve knowledge of nature. This becomes a problem for Kuhn since knowledge is typically defined as entailing truth, as the traditional analysis of knowledge maintains that 'S knows p iff (i) S believes that p, (ii) S is justified in believing that p, and (iii) *p is true*.' Under this analysis, knowledge is defined as 'justified true belief'. If this is the case, then when Kuhn maintains in G_1 that science achieves knowledge of nature, it follows that this claim entails the further claim that 'Science achieves justified beliefs about nature that are true'. But this claim directly contradicts G_2, which holds that science does not achieve the truth about nature.

Call this problem, the *problem of inconsistency*: Kuhn cannot consistently maintain, on the one hand, that science achieves knowledge of nature, and on the other, that science does not converge towards truth. Kuhn's double-goal is, thus, inconsistent, since truth is a necessary condition for knowledge. Thus, if Kuhn wishes to maintain his first goal, then he cannot, at the same time, uphold his second goal.

11.5 Defending Kuhn's Enterprise: Nature and Truth

I maintain that, while the problem of inconsistency threatens to undermine Kuhn's enterprise, there is a response that Kuhn can make which will help him to overcome the problem: namely, accept the claim that science achieves the truth about nature, with the qualification that a new alternative theory of truth is introduced. While this response deserves a focused and more developed explanation, I will explain here why it is a suitable response for Kuhn, discuss the steps that need to be taken in this response, and sketch the guidelines one would need to adhere to determine what theory of truth would be consistent with Kuhn's enterprise.

11.5.1 The Truth About Nature

My proposal that science can achieve truth (with a suitable theory of truth introduced) is plausible for Kuhn's enterprise in the sense that Kuhn, himself, believed that he needed a theory of truth. Particularly, he suggested that an account of truth "is badly needed to replace" CTT, and this account needs to "introduce minimal laws of logic (in particular, the law of non-contradiction) and make adhering to them a precondition for the rationality of evaluations. On this view, as I wish to employ it, the essential function of the concept of truth is to require choice between acceptance and rejection of a statement or a theory in the face of evidence shared by all" (Kuhn 1990, p. 99). This response can be effective only insofar as it can remain consistent with Kuhn's philosophy of science. This brings us to the steps that need to be taken to properly introduce an account of truth. The first step in this response focuses on what Kuhn means by "nature" in his double goal. On the one hand, his criticism of truth and separation of science from the pursuit of truth concerns looking at the independent world. Here, G_2 should be interpreted as assuming nature to be the independent world, and so science is not achieving the truth about nature as the independent world. Meanwhile, when Kuhn holds that science achieves knowledge of nature in G_1, nature here should be understood as the world of experience, partially determined by the given paradigm. Here, I follow Paul Hoyningen-Huene's use of the *phenomenal world* as a world that "changes over the course of a revolutionary transformation in science." (1993, p. 32). This interpretation leads to a slight reformulation of Kuhn's double goal:

- G_1: Science is cognitive; it achieves knowledge of the phenomenal world; the criteria it uses are thus epistemic'
- G_2: Scientific beliefs are not directed towards truth (exact or approximate) as it obtains in the independent world and the subject of truth claims is not a relation between belief and the independent world.

With this reformulation in mind, we can see that it may be the case that there really isn't a contradiction between G_1 and G_2 after all, given the ambiguity implicit in each goal of the double goal: science achieves knowledge of nature as the phenomenal world, but does not achieve the truth about nature as the independent world.

The second step in this response would then be to introduce a new theory of truth into Kuhn's enterprise. But what sort of truth are we working with here? We know what theory it cannot be: namely, the traditional account of CTT. Likewise, following Kuhn's criticisms of CTT, we can extrapolate features of a suitable account of truth. First, the Kuhnian theory of truth should not have reference to the independent world; rather it should concern itself only with the phenomenal world. Second, it should not require a neutral epistemic position that transcends the paradigm, but instead is grounded within the paradigm. Third, it is not an absolute notion of truth (in the sense that it is universal). On the contrary, it appears as though a Kuhnian truth should be more relative, but at the same time, remain objective within the paradigm.

11.5.2 The Nature of Truth

I argue that the most plausible candidate for an operational theory of truth in Kuhn's philosophy is an alternative version of the correspondence theory, one that pivots around a new epistemological thesis and appropriate metaphysical claims. I call this modified version, the *phenomenal-world correspondence theory of truth* (PCTT). In order to understand how to formulate PCTT, we will first need to examine Kuhn's epistemic criteria. Kuhn (1993) discusses two senses for characterizing certain criteria (e.g., simplicity) as epistemic. First, we can say that simplicity is an epistemic criterion in terms of CTT: "the sense in which the truth or falsity of a statement or theory is a function of its relation to a real world, independent of mind and culture." Meanwhile, the second sense for characterizing certain criteria as epistemic is based on Hans Reichenbach's distinction between two understandings of Kant's *a priori*. For Reichenbach, on the one hand, we can consider the Kantian a priori as that which "involves unrevisability and ... absolute fixity for all times." On the other hand, we can understand it as that which is "constitutive of the concept of the object of knowledge" (Kuhn 1993, p. 245). Kuhn points out that both ways of considering the Kantian a priori make up the second sense of epistemic criteria insofar as both meanings look, not towards the independent world, but towards the 'mind-dependent world'.

Now, Kuhn espouses the sense of epistemic criteria that follows Reichenbach's second interpretation of the Kantian a priori; that is, for Kuhn, the epistemic criteria for evaluating statements or theories for acceptance or rejection is one that is constitutive of the concept of the object of knowledge. As it is constitutive of the concept of the object, Kuhn's epistemic criteria concern the categories that are relative to time, place, and culture. In Kuhn's terminology, we can say that Kuhn's criteria concern the lexicons that are relative to a scientific paradigm.

So what sort of theory of truth fits consistently with Kuhn's dismissal of both CTT and the independent world, as well as within Kuhn's epistemic criteria concerning paradigms? I contend that Kuhn can and should resort to a correspondence theory of truth—the *phenomenal-world correspondence theory of truth* (PCTT). This theory will be formulated in such a way so as to maintain consistency with Kuhn's rejection of CTT (thereby showing that Kuhn *can* resort to a correspondence theory of truth). Furthermore, the use of PCTT will be sufficient for overcoming the problem of inconsistency (thereby showing that Kuhn *should* resort to a correspondence theory of truth).

PCTT has the same structure as CTT: it includes two metaphysical claims, one epistemic claim, and one claim concerning language. But the content of these claims will differ from CTT based upon a different metaphysical framework, where truth now operates within a world that is (partially) dependent upon human consciousness. Thus, under PCTT, we will have:

- P_1: Truth is a property of statements and propositions.
- $P_2{}^1$: A statement or a proposition is true if and only if it corresponds to a
- fact that obtains in the *phenomenal world*.
- $C_1{}^1$: We have direct access to the *phenomenal world* by experience and this access is reinforced by agreement among other people.
- $C_2{}^1$: Language has the capacity to represent the facts that obtain in the *phenomenal world*.

11.5.3   Calming Kuhn's Worries

Having presented the claims of the *phenomenal-world correspondence theory of truth* (PCTT), one may immediately question whether or not Kuhn can accept this theory of truth into his philosophy of science. After all, we have already seen that Kuhn would like to do away with the notion of truth in science and so would be resistant to the idea of embracing PCTT. So, before we move on to discuss whether or not Kuhn *should* adopt this theory of truth, it will be important to see whether or Kuhn *can* accept this theory, given his own inhibitions about correspondence theories of truth.

We can anticipate two distinct problems Kuhn would have with PCTT. The first problem concerns whether or not Kuhn would accept even an 'alternate' correspondence theory of truth, given Kuhn's straightforward rejection of the traditional correspondence theory. But the driving force behind Kuhn's rejection of this version is the theory's use of the 'independent world'. He tells us: that "what is fundamentally at stake is . . . the correspondence theory of truth, the notion that the goal, when evaluating scientific laws or theories, is to determine whether or not they correspond to *an external, mind-independent world* [emphasis mine]. It is that notion, whether in an absolute or probabilistic form, that I'm persuaded must vanish together with foundationalism" (Kuhn 1991, p. 95). It is thus the use of the independent world and its subsequent incorporation into science that leads Kuhn to not only attack the traditional correspondence theory (as we have seen in his attack against C_1), but to also campaign to remove the notion of truth altogether from science.

However, as we have seen, the *phenomenal-world correspondence theory*, or PCTT, does not rely solely upon the independent world. As we see from $P2^1$, this theory of truth does not claim that a statement is true if and only if it corresponds to a fact that *obtains in the independent world*; rather the emphasis is upon the world of experience—a world that is characterized not only by the independent world, but also influenced by one's consciousness, or the *phenomenal world*. By shifting the emphasis of truth onto the world of experience, as opposed to the external, mind-independent world, PCTT avoids the epistemic skepticism Kuhn holds toward the traditional correspondence theory, because there is no question of whether or not we have epistemic access to the world where facts obtain, since the phenomenal world is the world which we experience.

The second problem that concerns us is the question of whether or not Kuhn would accept the use of the phenomenal world into his metaphysics. While he dismisses talk of the independent world, Kuhn is also wary of using the notion of the dependent world (i.e., a phenomenal world). "The metaphor of a mind-dependent world. . . proves to be deeply misleading" because, Kuhn tells us, we allow for the possibility that the world is "an invention or construction of the creatures which inhabit it" (Kuhn 1990, p. 103). This possibility suggests that the world is under the conscientious control of the perceiver. Furthermore, the world in no way would be a given—the world is merely malleable and alterable by the perceiver, as the artist can alter her painting. As he is not an idealist, Kuhn maintains that the mind-dependent world as an invention is not the type of world he considers in his philosophy of science. We "find the world already in place, its rudiments at their birth and its increasingly full actuality during their educational socialization." Nor is the world malleable by the mind of the perceiver; it is stable, regardless of what the perceiver wants the world to be like: "it is entirely solid: not in the least respectful of the observer's wishes and desires; quite capable of providing decisive evidence against invented hypotheses which fail to match its behavior. Creatures born into it must take it as they find it" (Kuhn 1991, p. 101).

Kuhn, however, does not have to worry about this misleading conception of the mind-dependent world-as -invention in PCTT. First, the world within this theory is not a world of invention. It is not a world conscientiously created by the human being or any other creature—what makes it a partially 'mind-dependent' world is that it is a world that depends upon the lexicons that filter the world for the perceiver.[2] The world, then, is prior to the perceiver's experience of it insofar as the perceiver is born into a world. Thus, the perceiver does not willfully create the phenomenal world. Likewise, the perceiver cannot change the world any way she likes. The phenomenal world is not malleable and open to change on any whim. In this sense, then, the phenomenal world is solid and stable. Thus, the world in PCTT remains consistent with Kuhn's concerns about the dependent world, since it is not a world that is controlled by human awareness and invention.[3]

11.6 Conclusion

Kuhn's philosophical enterprise currently faces the problem of inconsistency, an internal contradiction insofar as Kuhn maintains that science achieves knowledge of nature, but, all the while, fails to achieve the truth about nature. I have shown that Kuhn's philosophy of science can overcome this problem by accepting that science achieves both knowledge and truth about nature. This solution entails a slight revision of Kuhn's double-goal at the heart of his enterprise, and an introduction of a new theory of truth into Kuhn's philosophy. I maintain that the most plausible candidate is a revised version of the correspondence theory of truth, one

---

[2] Kuhn defines a lexicon as "a conceptual scheme, where the 'very notion' of a conceptual scheme is not that of a set of beliefs but of a particular operating mode of a mental module prerequisite to having beliefs, a mode that at once supplies and bounds the set of beliefs it is possible to conceive" (1991, p. 94).

[3] Kuhn, himself, seemed to have been growing more comfortable accepting such discussion about a mind-dependent, or phenomenal world. In Kuhn's (1990), he discusses the co-dependent world as a world that is constitutive of intentionality and mental representations, which he traces back to his original emphasis in *Structure* as his "recourse to gestalt switches, seeing as understanding, and so on." And while Kuhn suggests that he was wary of using such manners of discussing world change, by 1991, he seemed more confident in discussing his metaphysical picture of changing worlds so that he could now return to references to his gestalt switch analogies and empirical structures of lexicons with a clearer understanding of what he had in mind in *Structure*.

that is able to incorporate the phenomenal worlds that Kuhn stressed to be either paradigm- or lexicon-dependent. By eliminating the notion of correspondence to a mind-independent world (which was Kuhn's initial reason for rejecting the traditional correspondence theory), the *phenomenal-world correspondence theory* of truth is able to help resolve the tension within Kuhn's enterprise and not disrupt the central philosophical claims he argued for in *Structure* onward.

References

Hoyningen-Huene, P. 1993. *Reconstructing scientific revolutions*. Chicago: University of Chicago Press.

Kuhn, T. 1962/1970. *The structure of scientific revolutions*. Chicago: University of Chicago Press.

Kuhn, T. 1990. The road since structure. In *PSA 1990. Proceedings of the 1990 Biennial Meeting of the Philosophy of Science Association*, vol. 2, ed. A. Fine, M. Forbes, and L. Wessels, 3–13. East Lansing: Philosophy of Science Association.

Kuhn, T. 1991. The trouble with the historical philosophy of science. Robert and Maurine Rothschild Distinguished Lecture presented on November 19, 1991, and published by Harvard University, Cambridge, MA in 1992. Reprinted in Kuhn (2000), 105–120.

Kuhn, T. 1993. Afterwords. In *World changes. Thomas Kuhn and the nature of science*, ed. P. Horwich, 311–341. Cambridge: MIT Press.

Kuhn, T. 2000. *The road since structure*. ed. J. Conant and J. Haugeland. Chicago: University of Chicago Press.

Putnam, H. 1981. *Reason, truth, and history*. Cambridge: Cambridge University Press.

Sankey, H. 1993. Kuhn's changing concept of incommensurability. *British Journal of the Philosophy of Science* 44:749–774.

Chapter 12
Kuhn's Social Epistemology and the Sociology of Science

K. Brad Wray

12.1 Introduction

Thomas Kuhn's *Structure* has had a profound effect on both the philosophy of science and the sociology of science, but in very different ways. Kuhn was in fact surprised by the book's reception, as well as its subsequent appeal. He did not and could not have anticipated the way it would be read by sociologists and philosophers of science.

On the one hand, Kuhn's intended audience was philosophers of science. But philosophers generally responded quite critically to the book. They saw it as a threat to the rationality of science. On the other hand, sociologists of science found the book to be insightful and stimulating. Like no one else before, Kuhn drew attention to the culture of science, specifically, the fact that scientists work in research traditions (see Barnes 1974, p. 48). *Structure* set sociologists of science in new directions. Sociologists began to study how social factors influence the content of science. Kuhn was disappointed with both of these responses, and he spent much of the remainder of his career trying to clarify his intentions, distinguishing his own project from the new sociology of science inspired by *Structure*, and arguing for the relevance of his work to philosophy of science.

In this chapter, I aim to clarify the relationship between Kuhn's social epistemology of science and the sociology of science. Moreover, I aim to clarify what Kuhn's positive legacy is to the philosophy of science. Kuhn, I argue, has provided us with the foundations for building a *social epistemology of science*.

I begin by recounting Kuhn's relationship to the sociology of science. First, I examine the influence of sociology of science on *Structure*. Second, I examine early responses to Kuhn's work by sociologists of science. Third, I examine Kuhn's views on the relevance of sociology to his own work. Finally, I examine his constructive contributions to the epistemology of science.

K. B. Wray (✉)
Department of Philosophy, State University of New York,
Oswego 212 Campus Center, 13126, Oswego, NY, USA
e-mail: brad.wray@oswego.edu

© Springer International Publishing Switzerland 2015   167
W. J. Devlin, A. Bokulich (eds.), *Kuhn's Structure of Scientific Revolutions—50 Years On*,
Boston Studies in the Philosophy and History of Science 311,
DOI 10.1007/978-3-319-13383-6_12

Table 12.1 Percentage of citations in *Structure* by subject field

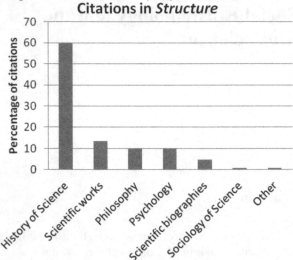

12.2 The Structure of *Structure*: A Citation Analysis

In order to understand the relationship between the sociology of science and Kuhn's project, it is worth examining what influenced Kuhn when he wrote *Structure*. Clearly, an author is influenced by more sources than he or she cites, but citations are a good place to start looking in order to understand the influences on an author.[1]

Kuhn cites 127 different sources in the first edition of *Structure*, with a total of 206 citations. An analysis of these sources suggests that, even though *Structure* profoundly influenced scholarship in history, philosophy, and sociology of science, Kuhn drew mostly on work in the history of science (Table 12.1).

In fact, 60 % of the sources cited in *Structure* are works in the history of science (that is, 76/127 sources). An additional 5 % of the sources cited are scientific biographies (that is, 6/127 sources). And an additional 13 % of the sources are scientific works, either contemporary scientific articles or historical classics, like Darwin's *On the Origins of Species* and Galileo's *Dialogue on Two New Sciences* (that is, 17/127 sources).

Nine of the ten most frequently cited sources in *Structure* are sources in the history of science. And these nine sources account for 25 % of the citations in *Structure*.[2]

---

[1] Kuhn does not cite some of the sources that clearly influenced him. For example, though he mentions Ludwik Fleck's *Entstehung und Entwicklung einer wissenschaftlichen Tatsache* in the Preface of *Structure*, he does not cite it (see Kuhn 1969, pp. viii–ix). Nor does he cite Hans Reichenbach's *Experience and Prediction*, the book which initially led him to Fleck's book (see Kuhn 1979, p. viii).

[2] I. B. Cohen's *Franklin and Newton* and E. T. Whittaker's *A History of the Theories of Aether and Electricity* are each cited eight times. Kuhn's own *Copernican Revolution* is cited seven times.

Given that most of Kuhn's own publications before *Structure* were in the history of science, it should not surprise us that most of the citations in *Structure* were to sources in that field.[3]

Despite the fact that the bulk of references in *Structure* are to sources in the history of science, Kuhn thought of *Structure* as a contribution to the philosophy of science (see Kuhn 1997, p. 276). But surprisingly there are few references to philosophical sources in *Structure*. In fact, Kuhn cites only 13 philosophical sources (a mere 10 % of the total sources cited). The philosophical sources cited include: Wittgenstein's *Philosophical Investigations*, Nelson Goodman's *Structure of Appearance*, Quine's "Two Dogmas of Empiricism," and Ernst Gombrich's *Art and Illusion*.[4] The three most important sources in the philosophy of science that Kuhn cites are Karl Popper's *Logic of Scientific Discovery*, Ernest Nagel's contribution to the positivists' *International Encyclopedia of Unified Science*, *Principles of the Theory of Probability*, and Norwood Russell Hanson's *Patterns of Discovery*.[5] Nagel and Popper are cited as representative proponents of two views of testing in science, views that Kuhn regards as mistaken (see Kuhn 1962, Chap. 12). Nagel represents probabilistic verificationism, and Popper represents falsificationism. Contrary to what Popper and Nagel suggest, Kuhn insists that theories are not merely tested against nature. Rather, Kuhn claims that "testing occurs as part of the competition between *two rival paradigms* for the allegiance of the scientific community" (Kuhn 1962, p. 144; emphasis added). Testing is essentially comparative. This is one of the key lessons Kuhn sought to teach philosophers of science. And his evidence for this claim comes from the history of science.[6]

Roller and Roller's *The Development of the Concept of Electric Charge* and J. R. Partington's *A Short History of Chemistry* are each cited six times. H. Metzger's *Newton, Stahl, Boerhaave et la doctrine chimique* is cited five times. W. Whewell's *History of the Inductive Sciences*, S. P. Thompson's *Life of William Thompson Baron Kelvin of Largs*, and A. Koyré's *Etudes Galiléennes* are each cited four times. The other source in the list of the ten most cited sources is N. R. Hanson's *Patterns of Discovery*. It is cited four times.

[3] Kuhn's formal training, from his bachelor's degree to his doctoral degree, was in physics. Kuhn learned the history of science working with his mentor, J. B. Conant, who designed and initially taught the General Education science courses at Harvard. Kuhn's time as a Harvard fellow was spent retooling for this new career.

Stephen Brush (2000) has examined the impact of Kuhn's historical research. Remarkably Kuhn had little impact in that field. His historical articles and books are seldom cited by historians (see Brush 2000, Table 1, p. 45). In addition, Kuhn's key claims in *Black Body Theory and the Quantum Discontinuity: 1894–1912* have not been integrated into physics textbooks (see Brush 2000, pp. 52–54).

[4] Two of the references are to obscure pieces from the journal *Philosophy of Science* written by scientists, an article by James Senior (see Kuhn 1962, p. 50), and a book review of Kuhn's *Copernican Revolution* by P. P. Wiener (see Kuhn 1962, p. 97).

[5] Hanson is the only philosopher to receive more than *two* citations. Kuhn, though, thanks Ernest Nagel in the Preface, along with Paul Feyerabend, and John Heilbron (Kuhn 1962, p. xiv).

[6] There is a dearth of citations to the work of the positivists in *Structure*, even though Kuhn explicitly criticizes positivism (see Kuhn 1962, p. 146). Kuhn confesses that when he was writing *Structure* he was quite ignorant of contemporary work by the positivists, including that of Carnap (see Kuhn 1997, pp. 305–306). Kuhn describes his target as "that sort of everyday image of logical positivism"

That Kuhn cites so many sources from the history of science in a book that purports to be a contribution to the philosophy of science may seem to undermine his expressed intentions. But Kuhn explains in the introductory chapter of *Structure* that he means to pursue his philosophical study of science in a different manner than traditional philosophical studies of science. Kuhn sought neither an idealized account of science, nor a logical analysis of the relations between theory and data. Rather, he proposed to study the history of science to ascertain how science *really* works. Kuhn thought of the history of science as "a source of phenomena to which theories about knowledge" must apply (Kuhn 1962, p. 9). That is, the history of science would supply the data that would constrain our theorizing about science. Alexander Bird notes that Kuhn was a pioneer in naturalizing the epistemology of science (Bird 2000).

Interestingly, Kuhn cites as many sources in psychology as he does in philosophy (13 or 10%). About half of the sources in psychology that he cites are articles reporting research on perception that has some bearing on the issue of the theory-ladenness of observation.[7] These sources play an integral role in Kuhn's naturalized philosophy of science. They provide insight into how scientists *really* see the world.

Even more startling than the limited use Kuhn makes of philosophical sources in *Structure* is the scarcity of citations to work in the sociology of science. Kuhn cites only *one* source in the sociology of science in the first edition of *Structure* (that is, less than 1%), Bernard Barber's "Resistance by Scientists to Scientific Discovery," a paper published in *Science* (see Kuhn 1962, p. 24).[8] With so few citations to sources in the sociology of science, it is somewhat surprising that the book influenced subsequent developments in that field so profoundly.[9]

Kuhn's Postscript to the second edition of *Structure*, published in 1970, contains citations to an additional 27 sources. Thirteen of these are to philosophical sources, a number of them to the papers in the conference proceedings edited by Imre Lakatos

---

(306). Nagel (1939) and Reichenbach (1938) are the principal sources for Kuhn's view of positivism. The clearest statement of Kuhn's conception of positivism is in his review of Joseph Ben-David's *The Scientist's Role in Society*. Kuhn claims that "Ben-David emerges as an unregenerate positivist, a man who believes that *scientific ideas are ... responses to logic and experiment alone*" (Kuhn 1972, p. 168, emphasis added). Alan Richardson (2007) has discussed the idiosyncratic nature of Kuhn's account of positivism.

[7] The following sources cited in *Structure* report research findings that have a bearing on how the human mind works: Piaget's *The Child's Conception of Causality*, and *Les notions de mouvement et de vitesse chez l'enfant*, Bruner and Postman's "On the Perception of Incongruity: A Paradigm," Stratton's "Vision without Inversion of the Retinal Image," Carr's *An Introduction to Space Perception*, Hastorf's "The Influence of Suggestion on the Relationship between Stimulus Size and Perceived Distance," and Bruner et al.'s "Expectations and the Perception of Color."

[8] Kuhn also cites Harvey Lehman's *Age and Achievement*, which includes an influential analysis of the changing productivity of scientists as they age (see Kuhn 1962, p. 90). Though Lehman was a psychologist, his book was later criticized by sociologists, specifically Merton's students. I thank Dean Simonton for insightful background on Lehman.

[9] Sociology of science was in its infancy in 1962 when Kuhn initially published *Structure* (see Cole and Zuckerman 1975). Even as late as 1968, Kuhn described the sociology of science as underdeveloped (Kuhn 1968/1977, p. 121).

and Alan Musgrave that resulted from Kuhn's encounter with Popper in London in 1965. Seven of the 27 sources cited in the Postscript are to cutting edge work in the sociology of science, specifically research concerned with identifying the structure of scientific specialty communities, or "invisible colleges." Kuhn cites work by Warren Hagstrom, Derek Price and Don Beaver, Diana Crane, and Nick Mullins (see Kuhn 1996, p. 176, 178). Kuhn also cites Eugene Garfield's work (see Kuhn 1996, p. 178). Garfield was instrumental in the creation of the Science Citation Indices (SCI) which sociologists used in an effort to better understand the structure of scientific research communities. Kuhn thought that we could determine the membership of research communities by taking account of such things as "attendance at special conferences ... the distribution of draft manuscripts or galley proofs ... and ... formal and informal communication networks" (Kuhn 1996, pp. 177–178). This he regarded as a sociological project, but one that was relevant to developing an adequate epistemology of science.

12.3 Sociological Responses to *Structure*

Even before the publication of *Structure*, Kuhn both interacted with sociologists of science and was identified as a person who might have insights into a sociological study of science. Writing on behalf of *The Institute for the Unity of Science* in 1952, Philipp Frank asked Kuhn to be part of "a research project under the general title 'sociology of science'" (see Thomas S. Kuhn Papers, MC 240, Box 25).[10] Again, following a conference Kuhn attended on the History of Quantification, he was asked in 1959 by the President of the *Social Science Research Council*, Pendleton Herring, for "suggestions with regard to the further development of a sociological approach to the history of science" (see Thomas S. Kuhn Papers, MC 240, Box 23).

Kuhn corresponded with Robert Merton in the late 1950s, discussing both Kuhn's own paper on measurement, and Merton's paper "Priorities in Scientific Discovery" (see Thomas S. Kuhn Papers, MC 240, Box 22). Merton was especially struck by Kuhn's discussion of textbook science, a theme that Kuhn would return to in *Structure*, and one that has important sociological implications for understanding the culture of science. Textbook science not only leaves scientists with a distorted sense of the history of their disciplines, it also supports the view that the growth of science is cumulative, a view Kuhn aggressively sought to undermine.

Merton was not the only sociologist of science with whom Kuhn interacted during this period. Barber gave Kuhn feedback on a draft manuscript of *Structure* (see

---

[10] The Institute would later evolve into the Boston Colloquium for Philosophy of Science (see Isaac 2012, p. 234). I thank Alisa Bokulich for alerting me to this connection. George Reisch (2005) argues that though the Institute for the Unity of Science was intended to continue the work of the Vienna Circle, Frank's vision for the Institute was at odds with the direction that philosophy of science was going in America, a direction toward greater professionalization (see Reisch 2005, pp. 294–306).

Thomas S. Kuhn Papers, MC 240, Box 25). Barber was Merton's student, and, as noted above, Barber was the only sociologist Kuhn cites in *Structure*.

Thus, long before the formation of the Strong Programme, Kuhn's work resonated with sociologists of science. In a letter to Kuhn from 1962, Merton wrote that "more than any other *historian* of science I know, you combine a penetrating sense of scientists at work, of patterns of historical development, and of *sociological* processes in that development" (cited in Cole and Zuckerman 1975, p. 159; emphasis added). Even though Merton identified Kuhn as a *historian* of science, he found his work to be insightful from a sociological point of view.

Kuhn's impact on the sociology of science continued to be felt in the 1970s. In an article published in the mid-1970s on the emergence of the sociology of science as a scientific specialty, Cole and Zuckerman (1975) provide lists of the most cited authors in the sociology of science between 1950 and 1973. They list the data in five year periods: Kuhn ranks in the top 15 in the period from 1950 to 1954, that is, even before the publication of *Structure*; he ranks in the top 10 in the period from 1960 to 1964; and he ranks in the top 20 for the period from 1965 to 1969. And in the four year period from 1970 to 1973 Kuhn ranks in the top 10 (Table 12.2).[11]

When Cole and Zuckerman wrote this paper, they noted that "there are no signs that … Kuhn [is] … turning away from [his] interest in the sociology of science" (1975, p. 154). Thus, in the mid-1970s Kuhn was identified as an important sociologist of science by sociologists of science.

With the formation of the Strong Programme, Kuhn's influence on the sociology of science became even more pronounced. Kuhn served on the editorial board for *Science Studies*, a key publishing venue for the proponents of the Strong Programme, from the journal's inception in 1971 until 1976, when it was renamed *Social Studies of Science* (see Thomas S. Kuhn Papers, MC 240, Box 12). The editor, David Edge, initially asked Kuhn to serve on the editorial board because he anticipated that contributing authors, especially sociologists of science, would draw on Kuhn's theoretical framework (see letter dated 13th Nov. 1967, Thomas S. Kuhn Papers, MC 240, Box 12). Indeed, in 1976, when Kuhn insisted that he should no longer serve on the editorial board for the journal, Kuhn noted that he "liked the journal from the start, and [he] thinks it has been improving steadily and serving an increasingly significant function as it does so" (see Thomas S. Kuhn Papers, MC 240, Box 12). Thus, initially, Kuhn was somewhat supportive of and enthusiastic about the new research in the sociology of science.

Structure also played a key role in the curriculum for the Science Studies program at Edinburgh University, the principal training ground for the Strong Programme. *Structure* was the central text in a course titled "A Philosophical Approach to Science" (Bloor 1975, p. 507). No wonder the proponents of the Strong Programme

---

[11] Cole and Zuckerman examine publications and citations in nine journals: *American Sociological Review*, *American Journal of Sociology*, *Social Forces*, *Social Problems*, *Sociology of Education*, *British Journal of Sociology*, *Minerva*, *American Behavioral Scientist*, and *Science* (see Cole and Zuckerman 1975, p. 169, Note 27).

Table 12.2 Kuhn's influence in the sociology of science. (Based on Cole and Zuckerman 1975, p. 154)

	1950–1954	1960–1964	1965–1969	1970–1973
1.	Gilfillan	Merton	Merton	Merton
2.	Lundberg	Crombie	Price	Price
3.	Dewey	Barber	Garfield	Hagstrom
4.	Hart	Gillispie	Hagstrom	Cole, JR
5.	Parsons	Lazersfeld	Zuckerman	Ben-David
6.	Merton	Kornhauser	Gordon	Cole, S
7.	Weber	Flexner	Glaser	Zuckerman
8.	Shils	Goodrich	Garvey	Gaston
9.	Conant	*Kuhn*	Kessler	*Kuhn*
10.	Leighton	Caplow	Cartter	Crane
11.	Isard	Shepard	Ben-David	Barber
12.	Kautsky	Shryock	Barber	Cartter
13.	Lerner	Wilson	Pelz	Glaser
14.	Lasswell	Glaser	Cole, S	Ogburn
15.	*Kuhn*	Gilfillan	Cole, JR	McGee
16.			Gamson	
17.			Kaplan	
18.			Storer	
19.			Lazersfeld	
20.			*Kuhn*	

Kuhn did not rank among the top 20 sociologists in the period from 1955 to 1959

identified themselves as Kuhnians. But the Strong Programme took liberties with Kuhn's work. Attempting to draw out the logical consequences of his view, they ended up developing a position that Kuhn did not recognize as his own.

The Strong Programme was especially influenced by Kuhn's theory of concept learning and concept application, in particular, Kuhn's analysis of the child learning a variety of bird concepts (Kuhn [1974] 1977, pp. 293–319). Kuhn's account of concept application is the foundation of the Strong Programme's finitism (see Barnes et al. 1996, Chap. 3).[12] According to the Strong Programme's finitism, every act of classification is underdetermined by logic and evidence (see Barnes et al. 1996, p. 54). As a result, at every instance decisions must be made about how to apply concepts, and what is to count as an instance of the class or kind designated by a particular

---

[12] Kuhn's example of the child learning various bird concepts is discussed at length in Barnes (1982), and Barnes et al. (1996, pp. 49–53).

concept. The open-endedness of concept application makes room for the influence of social factors, like interests and relations of power. Hence, in this way the *content* of scientific theories became a legitimate subject of sociological investigation.

Strictly speaking, finitism should not be identified with Kuhn's own view of concept application. Rather, finitism is best described as *based* on a particular reading of Kuhn's view.[13] I have argued elsewhere that finitism is a far more radical form of nominalism than Kuhn accepts. Whereas the Strong Programme regards *every* act of classification as underdetermined, Kuhn merely insists that there is no unique way to classify things in the world (see Wray 2011, pp. 145–146). Thus, from Kuhn's perspective, the proponents of the Strong Programme exaggerate the open-endedness of concept application.

As far as Kuhn is concerned, in periods of normal science concept application is generally unproblematic. Working in a normal scientific tradition, scientists take for granted that the concepts supplied by the accepted theory are adequate for representing the structure of the world. It is only when anomalies are encountered that problems arise. Indeed, for Kuhn, an anomaly just *is* something that cannot be classified given the conceptual resources supplied by the accepted theory. It is when scientists try to resolve persistent anomalies that the underdetermination of concept application becomes important. In these situations scientists must often take an active role in determining how a particular concept is to be applied. Indeed, in such situations disputes often arise over what is the proper scope of a concept.

12.4   Kuhn and the Sociology of Science

Even though it was a philosophy of science that Kuhn aimed to develop in *Structure*, sociology of science was central to his project.[14] Kuhn claims that it was Ludwik Fleck's *Genesis and Development of a Scientific Fact* that "helped [him] to realize that the problems which concerned [him] had a fundamental *sociological* dimension" (Kuhn 1979, viii; emphasis added; see also Kuhn 1962, pp. viii–ix).[15] Kuhn continued to describe his own approach to the study of science as "quasi-sociological" (Kuhn 1997, p. 310). In the Preface to *The Essential Tension* Kuhn claims that his own work is "deeply sociological, but not in a way that permits that subject to be

---

[13] See Barnes et al. 1996, p. 207 Note 1 and Wray 2011, Chap. 9.

[14] In a letter to Charles Morris, dated July 31, 1953, where Kuhn describes the contents of his planned book, tentatively titled *The Structure of Scientific Revolutions*, Kuhn notes that his "basic problem is sociological, since … any theory which lasts must be embedded in [a] professional group" (see Thomas S. Kuhn Papers, MC 240, Box 25). I thank George Reisch for drawing my attention to this letter.

[15] In *Structure* Kuhn remarks in a footnote that his own work overlaps with Warren Hagstrom's work in the sociology of science. Hagstrom completed a Ph.D. in sociology at the University of California, Berkeley.

separated from epistemology" (1977b, p. xx).[16] He claims that "scientific knowledge is intrinsically a *group* product" (xx; emphasis in original). Kuhn elaborated on these themes when he received the J. D. Bernal Award from the Society for the Social Studies of Science (4S). Kuhn claimed that

> *Structure* is sociological in that it emphasizes the existence of scientific communities, insists that they be viewed as the producers of a special product, scientific knowledge, and suggests that the nature of that product can be understood in terms of what is special in the training and values of those groups. (Kuhn 1983, p. 28)

Kuhn confesses that when he wrote *Structure* he knew very little about sociology. In fact, he claims that "[he] proceeded to make up the sociology of such communities as [he] went along... [drawing] it from [his] experience with the interpretation of scientific texts supplemented by [his] experience as a student of physics" (28). Kuhn claims that his guiding question was: "why [has] the special nature of group practices in the sciences been so strikingly successful in resolving the problems scientists choose"? (28).[17]

Kuhn, though, was deeply troubled by the developments in the sociology of science initiated by the Strong Programme.[18] His concerns about the Strong Programme changed, influenced both by his own deepening understanding of *their* project, and in response to the evolution of the Strong Programme. It is worth distinguishing three separate criticisms Kuhn came to level against the Strong Programme.

Initially, Kuhn was concerned that the proponents of the Strong Programme misunderstood the role that values play in science. He contrasted their view with Merton's, a view that he regarded as essentially correct. Whereas Merton believed that science was characterized by a set of values which are more or less stable throughout the history of modern science, the Strong Programme denied this.[19] According to Kuhn, the proponents of the Strong Programme believe that "values vary from community to community and from time to time" (Kuhn 1977b, p. xxi). The Strong Programme's detailed historical studies emphasize the *contingencies* that influence the resolution of disputes in science. These studies purport to show that there is no set of values constitutive of *all* science. Kuhn disagreed with this view of science, insisting that the shared values constitutive of science play an important role in enabling scientists to realize their epistemic goals (1977b, pp. xxi–xxii).

---

[16] Incidentally, Hans Reichenbach had a similar view about the relevance of sociology to epistemology. He claimed that "the first task of epistemology [is its] *descriptive task* ... Epistemology in this respect is part of sociology" ([1938] 2006, p. 3).

[17] In a letter to the President of the *Social Science Research Council*, dated 21 December 1959, Kuhn claims that sociological studies of science were crucial to developing a more accurate account of science (see Thomas S. Kuhn Papers, MC 240, Box 23).

[18] Interestingly, Merton was alarmed about the Strong Programme before Kuhn was. In a letter to Kuhn dated 18 February 1976, Merton expressed concern about the "Kuhnians" and an alleged "Kuhn-vs-Merton" dispute that Merton believed the self-proclaimed Kuhnians were manufacturing (see Thomas Kuhn Papers, MC 240, Box 22).

[19] To some extent, the core values identified by Merton are the sociological equivalent of Popper's demarcation principle, intended to distinguish genuine science from pseudo-science.

By the time Kuhn was being honored by the Society for the Social Studies of Science he had a different concern. He was concerned that the Strong Programme's sociological studies of science, which focused "pre-dominantly [on] socio-economic interests," mistakenly "excluded the special cognitive interests inculcated by scientific training," like "love of truth... [and] fascination with puzzle solving" (Kuhn 1983, p. 30). Kuhn thus felt that proper sociological studies of the content of science must attend to the specific interests that motivate scientists, that is, their cognitive interests. Unfortunately, neither sociologists nor philosophers picked up on this concern.

Later, Kuhn raised a third criticism against the Strong Programme. He complained that the Strong Programme's studies of science "leave out the role of [nature]. Some of these people simply claim that it doesn't have any [role], that nobody has shown that it makes a difference" (Kuhn 1997, p. 317). Though Kuhn acknowledged in the 1990s that the proponents of the Strong Programme no longer held this extreme view he believed they failed to clarify what role nature plays in resolving disputes in science (Kuhn 1997, p. 317). Indeed, there is some basis for Kuhn's reading of the Strong Programme's position. In an explanation of how concepts are applied in *T. S. Kuhn and Social Science*, Barry Barnes claims that "*nature* sets *no constraints* on the form of the routine which is produced" (Barnes 1982, p. 29, emphasis added). Elaborating, Barnes explains that "any way of developing the accepted usage of a concept could equally well be agreed upon, since any application of a concept to an instance can be made out as correct and justified by the invocation of an appropriate weighting of similarity against difference" (29).

Kuhn, though, insists that nature plays a significant role in shaping scientists' beliefs. The world is "not in the least respectful of an observer's wishes and desires; quite capable of providing decisive evidence against invented hypotheses which fail to match its behavior" (Kuhn 1991, p. 101). Kuhn believed that an adequate understanding of science required attention to the role nature plays in resolving disputes. It is in this respect that Kuhn thought of his view as fundamentally different from the view of the Strong Programme.

Despite Kuhn's attitude toward the Strong Programme, many philosophers saw strong parallels between his view and the view of the Strong Programme. Kuhn's remarks at the end of *Structure* about how appeals to truth explain very little in science *may* sound like an anticipation of the Strong Programme's symmetry principle (see Kuhn 1962, pp. 169–170). But this was not Kuhn's intention.

The Strong Programme's symmetry principle is a methodological principle instructing sociologists of science to seek explanations for both true beliefs and false beliefs in terms of social causes (Barnes and Bloor 1982). In invoking this principle, the proponents of the Strong Programme want us to resist the temptation of thinking that true beliefs are caused by our successful interaction with nature, and only false beliefs have social causes. The symmetry principle is thus a corrective to reducing the sociology of science to the sociology of error.[20] Proponents of the Strong Programme insist that even true beliefs have social causes. Further, they insist that social causes of belief are not necessarily distorting.

Like the proponents of the Strong Programme, Kuhn believes that scientists are influenced by a variety of social factors. But, unlike the proponents of the Strong Programme, Kuhn is an internalist. He does not believe that factors external to science determine the outcome of scientific disputes.[21] Kuhn acknowledges that when the available evidence does not unequivocally support one of the competing hypotheses, scientists are influenced in their decision making by various subjective factors (see Kuhn 1977c). For example, Kepler's neo-Platonism led him to accept the Copernican theory more than a decade before Galileo's telescopic evidence was gathered (Kuhn 1977c, p. 323). According to Kuhn, such subjective factors ensure that there is an effective division of labor in the research community, and the viable competing hypotheses are developed. But Kuhn believes that when science is working well, controversies in science are resolved on the basis of a consideration of the epistemic merits of the competing views (see Wray 2011, Chap. 9).

It is Kuhn's internalism that both distinguishes his view from the view of the Strong Programme and makes his account a contribution to the *epistemology* of science. Disputes in science are ultimately resolved on the basis of a consideration of the available evidence. The proponents of the Strong Programme, on the other hand, reject the internalism/externalism distinction. Their point is that the sorts of factors that philosophers typically regard as external factors can play an integral role in determining the outcome of scientific disputes.[22]

12.5   Kuhn's Social Epistemology

I now want to turn to Kuhn's social epistemology. Specifically, I want to look at two ways in which the social structure of research communities figures in Kuhn's account of scientific change and helps us understand why science is so successful. One example is drawn from *Structure*, and the other example is drawn from Kuhn's later work, published in *The Road since Structure*.

---

[20] According to Laudan, "the sociology of knowledge may step in to explain beliefs if and only if those beliefs cannot be explained in terms of their rational merit" (Laudan 1977, p. 202).

[21] For further discussion of Kuhn and internalism, see Bird, Chap. 3, this volume; and for some challenges, see Mody, Chap. 7, this volume

[22] See Pinch (1979) and Barnes (1982, p. 118). I am not alone in insisting that Kuhn's project is rightly characterized as an epistemology of science. Robert Nola (2000) goes to great lengths to argue that the Strong Programme misappropriated Kuhn's work, and that Kuhn's concerns were continuous with the concerns of philosophers of science (see 89). Nola expresses Kuhn's view on the matter quite precisely in the following remark. "Even though sociology of *science* can play a role in the individuation of scientific communities, for Kuhn sociology of *scientific knowledge* plays very little role in theory choice and none in his account of the justification of his principles of theory choice" (Nola 2000, p. 80). Nola, though, believes that Kuhn's project ultimately fails.

First, consider the cyclical pattern of scientific change from normal science to crisis, from crisis to revolution, and from revolution to a new normal scientific tradition. In each of these phases of research, scientists encounter different sorts of challenges. And the structure of the research community changes in order to meet these challenges. Normal science is characterized by the uncritical acceptance of a theory in a research community. It is this uncritical acceptance of a theory that makes research during these periods so effective, giving the impression that the growth of scientific knowledge is cumulative, with no set-backs or Kuhn-loss. Scientists working in such a tradition have internalized the accepted norms, standards and concepts thoroughly. Consequently, they unreflectively make the sorts of discriminations between the phenomena that the accepted theory dictates. Indeed, working in a normal scientific tradition, scientists often *assume* that the concepts they use cut nature at the joints.

When the consensus that characterizes normal science breaks down, the research community is in a state of crisis. It is only then that scientists are willing to develop and seriously entertain alternative theories. The research community thus changes in an effort to meet the new challenge, a challenge that requires significant "retooling" (Kuhn 1996, p. 76). These changes in the structure of the research community through the cycle of normal science, crisis, and revolution are what enable scientists to make advances in the pursuit of their epistemic goals. This focus on the research community as the locus of scientific change is one of the key features that makes Kuhn's epistemology a *social* epistemology of science. Indeed, Kuhn would not have misled his readers if he had titled the book *The Structure of Scientific Research Communities*.

Structure ends as a bit of cliff-hanger. On the one hand, Kuhn insists that there is progress through changes of theory. The new theory enables scientists to explain things that were inexplicable before (see Kuhn 1962, p. 168). And after a change of theory, scientists are usually able to make more accurate predictions of the phenomena they study (see Kuhn 1962, p. 169). On the other hand, Kuhn claims that the more recently developed theory in a field is not aptly described as closer to the truth than the theory it replaced (Kuhn 1962, pp. 169–170; see also Kuhn [1992] 2000a, p. 115). This struck many philosophers as a denial of *real* progress in science. It also raises serious questions about how science can be increasing in precision despite the fact that scientists are not getting any closer to the truth. Nowhere in *Structure* do we get an alternative explanation for the increasing accuracy of theories over time.

In his later work, though, Kuhn begins to develop an alternative explanation. Again, Kuhn appeals to changes in the social structure of research communities in order to address problems in the epistemology of science. What Kuhn seeks to explain is how scientists are able to develop theories that are instrumentally superior to the theories they replace, despite the fact that these new theories may not be aptly described as closer to the truth than the theories they replace.

Kuhn came to believe that not all crises in science are resolved by a revolutionary change of theory. Some crises are resolved by dividing a field into two separate specialties, each responsible for a sub-set of the phenomena for which the parent field was originally responsible. Though each of the two specialties initially use a similar set of concepts inherited from the parent field, as scientists in each specialty devote more time to the research problems specific to their own specialty, the conceptual frameworks of the two specialties will begin to diverge. In time, there will be significant differences between the fields, differences that may even undermine effective communication between scientists in the two fields (Kuhn [1969] 1996, p. 177).

Elsewhere, I have illustrated this process with respect to the creation of two new scientific specialties: endocrinology and virology (Wray 2011, Chap. 7). Endocrinology was created as a new field when physiologists sought to explain phenomena that did not fit the accepted models of how the body works. Physiologists had assumed that the functions of the body were coordinated by the nervous system. Instead, the anomalous phenomena were explained in terms of the operation of chemical messengers, hormones, a new type of entity. The creation of virology as a scientific specialty, and its separation from bacteriology, followed a similar developmental path. Crucial to the creation of this new field was the development of a new concept, a virus, something that had properties that were fundamentally different from those of bacteria. Viruses cannot be reproduced in vitro, like living organisms. But unlike non-living toxic substances, viruses can replicate.[23]

Once a new scientific specialty is created, both the new specialty and the parent field from which it separates are able to make epistemic advances. Scientists working in the new field can focus their attention on developing concepts, instruments, and practices suited to studying the phenomena that fall within their field's domain. Inevitably, they will develop instruments and concepts that are of no use to those working in the parent field. And with the creation of the new specialty, scientists who continue to work in the parent field can relinquish responsibility for some previously unsolved problems and hitherto unexplained phenomena. Scientists in the parent field are thus no longer held back by the persistent anomalies that led to the creation of the new specialty in the first place. Not surprisingly, when scientists are concerned with a narrower range of phenomena, they are often able to develop models that enable them to achieve a greater degree of precision. These epistemic advances are achieved, in part, through a change in the social structure of science. In this respect, Kuhn's epistemology is aptly described as a social epistemology.

Incidentally, even in *Structure* Kuhn suggested that increasing specialization plays an important role in science (Kuhn 1962, p. 169). He regarded it is an inevitable *consequence* of the progress of science. Later, though, he came to regard specialization as the *cause* of the progress. It is because of specialization that scientists are able to explain as much as they can, and explain what they can with ever increasing precision.

A new picture of scientific change emerges in Kuhn's later work. The simple cycle of change presented in *Structure*, and discussed above, has been replaced. In mature fields, periods of normal science are interrupted in one of two ways. Either a crisis leads to a revolutionary change of theory, or anomalies that cause a crisis lead to the

---

[23] For a further discussion of this case see Wray (2011, Chap. 7).

Resolving Crises: Kuhn's Mature View

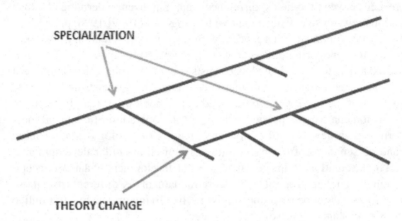

SPECIALIZATION

THEORY CHANGE

Fig. 12.1 Kuhn's mature view on the growth of science and how crises are resolved

creation of a new specialty which takes care of the anomalous phenomena, leaving the parent field more or less intact (Fig. 12.1).

12.6   Concluding Remarks

My aim in this chapter has been to clarify the nature of the relationship between Kuhn's project and the sociology of science. By now it should be clear how complicated the relationship is. Kuhn described his project as sociological in some respects, and he believed that philosophers could gain valuable insights about the structure of scientific research communities from sociologists of science. Developing a better understanding of the community structure of science was an integral part of his *epistemology* of science. Sociologists of science, both those working in the Mertonian tradition and those who followed, were profoundly influenced by Kuhn's work. They were excited by his claim that scientific research communities are cultures, with traditions that are resistant to change. And the Strong Programme felt that Kuhn gave them permission to investigate the content of science.

But Kuhn's relationship with sociologists of science became strained as the Strong Programme became the dominant view in the sociology of science, and as he fought to correct popular misconceptions about his view. Kuhn sought to show that his own project was philosophical, and thus different from the Strong Programme. Indeed, both the proponents of the Strong Programme and Kuhn attribute a significant role to the social structure of science in their explanations of scientific change. But, ultimately, Kuhn's concerns are *epistemic* in a traditional philosophical sense. Central to his project is a concern to explain how evidence and other epistemic considerations ultimately enable scientists to resolve their disputes. But unlike many philosophers of science, Kuhn does not think that we are warranted in claiming that we are getting ever closer to the truth. The history of science, marked by revolutionary changes of theory, cannot be reconciled with the view that we are getting ever closer to the truth. The success achieved in science is with respect to developing models and theories that are empirically successful. And some of these gains are made when scientists develop new specialties, and thus narrow the range of phenomena they seek to model.

Acknowledgments I thank Lori Nash for constructive feedback on numerous drafts of this paper. I also thank Kristina Rolin and George Reisch for critical feedback on earlier drafts of this paper. I thank Charles Hickey, my research assistant, for double-checking my citation counts in *Structure*. I thank my audiences at the following: the Philosophy Department Colloquium at the State University of New York, Oswego; the Boston Colloquium in Philosophy of Science; the Philosophy Department Colloquium at Rochester Institute of Technology; and the Orange Beach Epistemology Workshop sponsored by the University of South Alabama. I thank Evelyn Brister for inviting me to RIT, and Ted Poston of the University of South Alabama for inviting me to the Orange Beach Epistemology Workshop. I thank Alisa Bokulich, the Director of the Center for Philosophy and History of Science at Boston University, for inviting me to be part of the *Robert S. Cohen Forum* and Colloquium dedicated to the theme "50 Years since Kuhn's *Structure of Scientific Revolutions*." I thank Nora Murphy, the Archivist for Reference, Outreach and Instruction, and the rest of the staff at Massachusetts Institute of Technology, Institute Archives and Special Collections, for their assistance and for allowing me access to the Thomas S. Kuhn Papers. Finally, I thank SUNY-Oswego for supporting my research trip to MIT to visit the archives.

References

Barber, B. 1987. The emergence and maturation of the sociology of science. *Science & Technology Studies* 5 (3/4): 129–133.

Barnes, B. 1974. *Scientific knowledge and sociological theory*. London: Routledge and Kegan Paul.

Barnes, B. 1982. *T. S. Kuhn and social science*. New York: Columbia University Press.

Barnes, B., and D. Bloor. 1982. Relativism, rationalism and the sociology of knowledge. In *Rationality and relativism,* ed. M. Hollis and S. Lukes, 21–47. Cambridge: MIT Press.

Barnes, B., D. Bloor, and J. Henry. 1996. *Scientific knowledge: A sociological analysis*. Chicago: University of Chicago Press.

Bird, A. 2000. *Thomas Kuhn*. Princeton: Princeton University Press.

Bloor, D. 1975. Course bibliography: A philosophical approach to science. *Social Studies of Science* 5:507–517.

Brush, S. G. 2000. Thomas Kuhn as a historian of science. *Science & Education* 9:39–58.

Cole, J. R., and H. Zuckerman. 1975. The emergence of a scientific specialty: The self-exemplifying case of the sociology of science. In *The idea of social structure: Papers in honor of Robert K. Merton,* ed. L. A. Coser, 139–174. New York: Harcourt Brace Jovanovich.

Friedman, M. 2001. *Dynamics of reason: The 1999 Kant lectures at Stanford University*. Stanford: Center for the Study of Language and Information Publications.

Fuller, S. 2000. *Thomas Kuhn: A philosophical history for our times*. Chicago: University of Chicago Press.

Giere, R. 1997. Kuhn's legacy for North American philosophy of science. *Social Studies of Science* 27:496–498.

Isaac, J. 2012. *Working knowledge: Making the human sciences from Parsons to Kuhn*. Cambridge: Harvard University Press.

Kuhn, T. S. 1962. *The structure of scientific revolutions*. Chicago: University of Chicago Press.

Kuhn, T. S. 1968/1977. The history of science. In *The essential tension: Selected studies in scientific tradition and change*, ed. T. S. Kuhn, 105–126. Chicago: University of Chicago Press.

Kuhn, T. S. 1969/1996. Postscript—1969. In *The structure of scientific revolutions*, ed. T. S. Kuhn, 3rd ed., 174–210. Chicago: University of Chicago Press.

Kuhn, T. S. 1972. Scientific growth: Reflections on Ben-David's 'scientific role'. *Minerva* 10:166–178.

Kuhn, T. S. 1974/1977. Second thoughts on paradigms. *The essential tension: Selected studies in scientific tradition and change*, ed. T. S. Kuhn, 293–319. Chicago: University of Chicago Press.

Kuhn, T. S. 1977a. *The essential tension: Selected studies in scientific tradition and change*. Chicago: University of Chicago Press.

Kuhn, T. S. 1977b. Preface. *The essential tension: Selected studies in scientific tradition and change*, ed. T. S. Kuhn, ix–xxiii. Chicago: University of Chicago Press.

Kuhn, T. S. 1977c. Objectivity, value judgment and theory choice. *The essential tension: Selected studies in scientific tradition and change*, ed. T. S. Kuhn, 320–339. Chicago: University of Chicago Press.

Kuhn, T. S. 1979. Forward. In *Ludwig Fleck, Genesis and development of a scientific fact*, ed. T. J. Trenn and R. K. Merton, vii–xi. Trans. F. Bradley and T. J. Trenn. Chicago: University of Chicago Press.

Kuhn, T. S. 1983. Reflections on receiving the John Desmond Bernal award. *4S Review* 1 (4) 26–30.

Kuhn, T. S. 1991/2000. The road since structure. In *The road since structure: Philosophical essays, 1970–1993, with an autobiographical interview*. ed. T. S. Kuhn, J. Conant, and J. Haugeland, 90–104. Chicago: University of Chicago Press.

Kuhn, T. S. 1992/2000. The trouble with the historical philosophy of science. In *The road since structure: Philosophical essays, 1970–1993, with an autobiographical interview*, ed. T. S. Kuhn, J. Conant, and J. Haugeland, 105–120. Chicago: University of Chicago Press.

Kuhn, T. S. 1996. *Structure of scientific revolutions*. 3rd ed. Chicago: University of Chicago Press.

Kuhn, T. S., J. Conant, and J. Haugeland. 2000. *The road since structure: Philosophical essays, 1970–1993, with an autobiographical interview*. Chicago: University of Chicago Press.

Kuhn, T. S. [1997] 2000b. A discussion with Thomas S. Kuhn. In *The road since structure: Philosophical essays, 1970–1993, with an autobiographical interview*, ed. T. S. Kuhn, J. Conant, and J. Haugeland, 255–323. Chicago: University of Chicago Press.

Laudan, L. 1977. *Progress and its problems: Towards a theory of scientific growth*. Berkeley: University of California Press.

Lakatos, I. 1970. Falsification and the methodology of scientific research programmes. In *Criticism and the growth of knowledge: Proceedings of the International Colloquium in the Philosophy of Science, London 1965*, 4, reprinted with corrections, ed. I. Lakatos and A. Musgrave, 91–196. Cambridge: Cambridge University Press.

Nagel, E. 1939. *Principles of the theory of probability*, 1: 6, of *International Encyclopedia of Unified Science*. Chicago: University of Chicago Press.

Nola, R. 2000. Saving Kuhn from the sociologists of science. *Science & Education* 9:77–90.

Pinch, T. 1979. Paradigm lost? A review symposium. *Isis* 70:253, 437–440.

Reichenbach, H. [1938] 2006. *Experience and prediction: An analysis of the foundations and the structure of knowledge*. Notre Dame: University of Notre Dame Press.

Reisch, G. A. 2005. *How the Cold War transformed philosophy of science: To the icy slopes of logic*. Cambridge: Cambridge University Press.

Richardson, A. 2007. 'That sort of everyday image of logical positivism': Thomas Kuhn and the logical empiricist philosophy of science. In *The Cambridge companion to logical empiricism*, ed. A. W. Richardson and T. Uebel, 346–369. Cambridge: Cambridge University Press.

Wray, K. B. 2011. *Kuhn's evolutionary social epistemology*. Cambridge: Cambridge University Press.

Archival Sources

Thomas S. Kuhn Papers, MC240, Box 12. Massachusetts Institute of Technology, Institute Archives
    and Special Collections, Cambridge, Massachusetts.
Thomas S. Kuhn Papers, MC240, Box 22. Massachusetts Institute of Technology, Institute Archives
    and Special Collections, Cambridge, Massachusetts.
Thomas S. Kuhn Papers, MC240, Box 23. Massachusetts Institute of Technology, Institute Archives
    and Special Collections, Cambridge, Massachusetts.
Thomas S. Kuhn Papers, MC240, Box 25. Massachusetts Institute of Technology, Institute Archives
    and Special Collections, Cambridge, Massachusetts.

Chapter 13
Kuhn's Development Before and After *Structure*

Paul Hoyningen-Huene

13.1 Introduction

Thomas Kuhn's thinking was never at rest. This was, as far as I can see, mostly triggered by external influences: an immense flow of published reactions to his work together with an astonishing amount of letters informally addressing different aspects of his work. In addition, however, Kuhn himself saw the need to further develop his views in several directions. A really comprehensive account of these development remains to be written[1] and I am addressing only two time slices of the whole. The first one is a comparatively short period from April 1961 to spring of 1962, and the other one the last period of Kuhn's life from the early 1980's to about 1995. The first concerns the changes that Kuhn made to the penultimate draft of *The Structure of Scientific Revolutions*. In the other time period Kuhn worked on his last and, unfortunately, unfinished book, entitled *The Plurality of Worlds: An Evolutionary Theory of Scientific Development*.

13.2 Proto-Structure

Proto-Structure is the name I gave to the penultimate draft of Kuhn's *The Structure of Scientific Revolutions* (Kuhn [1962] (1970), henceforth *Structure*). *Proto-Structure* was probably finished in April 1961 and given or sent to several people

---

[1] Certain aspects of Kuhn's development until the late 1980s are dealt with in Hoyningen-Huene (1993).

P. Hoyningen-Huene (✉)
Institute of Philosophy and Center for Philosophy and Ethics of Science,
Leibniz Universität, Hannover, Germany
e-mail: hoyningen@ww.uni-hannover.de

© Springer International Publishing Switzerland 2015   185
W. J. Devlin, A. Bokulich (eds.), *Kuhn's Structure of Scientific Revolutions—50 Years On*,
Boston Studies in the Philosophy and History of Science 311,
DOI 10.1007/978-3-319-13383-6_13

inviting criticism.[2] On the basis of the reactions he received, Kuhn transformed *Proto-Structure* into *Structure*. The reactions he deemed most important were by James B. Conant, president of Harvard University; Leonard K. Nash, his former Harvard colleague; Stanley Cavell, his Berkeley colleague; Paul K. Feyerabend, his Berkeley colleague; Ernest Nagel, Columbia; H. Pierre Noyes, physicist at Lawrence Radiation Laboratory (one of Kuhn's roommates at Harvard); and John L. Heilbron, one of Kuhn's students.[3]

There are countless minor differences between *Proto-Structure* and *Structure*. However, there are also several significant differences. I shall focus on two of them, regarding Kuhn's novel concept of normal science and Kuhn's discussion of the well-entrenched distinction between the context of discovery and the context of justification.

13.2.1 Normal Science

Let us first look at how normal science is treated in *Structure*. There are two chapters dealing with normal science, Chaps. 5 and 6. The outline of Chap. 5 entitled "Normal science as puzzle-solving" is as follows. In §§ 1–2, Kuhn introduces normal science with the, at the time, surprising thesis that normal science does not strive for unexpected novelties. This was surprising because the reigning stereotype was that science is successful because it strives for, and indeed often produces, unexpected novelties. But if normal science does not strive for unexpected novelties, how can scientists be motivated to pursue such an enterprise? Kuhn's answer is that the motivation for normal science is the same as for "puzzle-solving" (§ 3). Again, this is a surprising answer because the analogy between a specific form of scientific practice and puzzles, such as chess puzzles, seems to be far-fetched. Kuhn therefore develops the parallels between puzzles and problems of normal science in §§ 4–13. These parallels include the existence of various rules for legitimate procedures and for the identification of acceptable solutions; in normal science these rules are derived from paradigms (§§ 7–13). Kuhn ends Chap. 5 with the somehow irritating remark that the expression "rules" is misleading in the given context (§ 14). This remark leads over to Chap. 6.

Chapter 6 of *Structure* is entitled "The priority of paradigms". Kuhn describes first that the historian discovers paradigms as the particular loci of commitment in normal science (§ 1). Rules of normal science, if existent, are derived from paradigms (§ 2). However, paradigms can guide science even without rules (§ 3). Two main questions regarding this statement emerge. First, what does this statement mean? This question is answered with recourse to Ludwig Wittgenstein's theory of concept use via family resemblance (§§ 4–6). The second question is whether the claimed absence of rules

---

[2] For the dating of *Proto-Structure*, see Hoyningen-Huene (2006b, p. 611).
[3] This is what Kuhn tells his readers in Kuhn ([1962] 1970), pp. xi–xii.

in normal science is plausible. Kuhn gives various reasons why the answer to this question is in the positive (§§ 7–12).

Already, this short sketch of the train of thought in Chaps. 5 and 6 of *Structure* reveals considerable tension between them. Whereas Chap. 5 develops the parallel of normal science with puzzle-solving in terms of rules, and takes "rules" somehow back at the end, Chap. 6 develops the thesis that paradigms guide science *without* rules. What does Chap. 6 make of Chap. 5? Why does Kuhn accept the tension between the two chapters instead of rewriting them? For a long time, my working hypothesis was that Kuhn needed the talk of rules in order to develop the parallel to puzzle-solving, which he deemed important. However, if normal science could be practiced without rules, he should have realized that the parallel simply broke down, despite its originality and potential conveying of insight. So, a puzzle remained. Perhaps this puzzle could be resolved by investigating the history of these two chapters in the stage immediately before *Structure*, i.e., in *Proto-Structure*.

Before going to the details of *Proto-Structure*, let us first compare the tables of contents of *Structure* and *Proto-Structure*. The Table of Contents of *Proto-Structure* is as follows:

I. Introduction
II. The Route to Normal Science
III. The Nature of Normal Science
IV. Normal Science as Rule-Determined
V. Anomaly and the Emergence of Scientific Discoveries
VI. Crisis as the Prelude to Scientific Theories
VII. The Response to Crisis
VIII. The Nature and Necessity of Scientific Revolutions
IX. Revolutions as Changes of World View
X. The Invisibility of Revolutions
XI. The Resolution of Revolutions
XII. Progress through Revolutions.

Apart from some minor differences to the Table of Contents of *Structure*, there is one major and completely surprising change. Where *Structure* features the two Chaps. 5 and 6 on "Normal Science as Puzzle-solving" and on "The Priority of Paradigms", *Proto-Structure* features only one chapter entitled "Normal Science as Rule-Determined". The following chapters of *Proto-Structure* bear (with the unimportant exception of Chap. 7) exactly the same titles as the corresponding chapters of *Structure*; only the chapter numbers are diminished by one as a consequence of the splitting of Chap. 5 of *Proto-Structure* into two chapters in *Structure*.

How does Kuhn deal with normal science in *Proto-Structure*? Already the title of Chap. 4 "Normal science as rule-determined" indicates that between *Proto-Structure* and *Structure* a major break occurred. Remember that, in *Structure*, Kuhn used rules only in a transitory fashion in order to develop the parallel of normal science with puzzles. After he had developed the parallel, he took rules back in order to develop one of the principal topics that made him and his book famous: the priory of paradigms! In *Proto-Structure*, however, nothing of this revolutionary idea seems to be present. Instead, at least as far as the title is concerned, the idea is presented that normal science is rule-directed, which seems to be nothing else than the rather old-fashioned idea that science is governed by the scientific method. And this is, indeed, what Kuhn does in the first half the chapter.

In the beginning of Chap. 5 of *Proto-Structure*, Kuhn claims (as in *Structure*) that normal science does not strive for unexpected novelties (§§ 1–2); instead, its motivation is "puzzle-solving" (§ 3). He then develops the parallel between puzzles and normal science: they are both practices that are based on (explicit or implicit) rules (§§ 3–6) (similar to *Structure*). However, in the following paragraphs 7–12, Kuhn takes partially back the role of rules by stating that normal science is *not* fully determined by rules. Instead, scientists practice their trade by working from paradigm examples. Then, the chapter abruptly ends—the following chapter is already on the role of anomalies. What remains entirely open is how exactly the guidance of normal science by paradigms works, as opposed to a guidance by rules. And, most surprisingly, in the light of *Structure*, *there is absolutely no Wittgenstein in this chapter of Proto-Structure!* More to the point, in *all* of Kuhn's writings up to 1961, there is *only one indirect reference* to Wittgenstein, namely when Kuhn claims that the early theories in electrical research "had no more than a family resemblance".[4] The missing Wittgenstein in Kuhn up to 1961 would be completely mysterious if the standard story about Kuhn's encounter with Wittgenstein were true: that Kuhn read Wittgenstein's *Philosophical Investigations* in 1959.[5] It is very implausible that Kuhn read Wittgenstein in 1959, worked on *Proto-Structure* until April 1961 without mentioning Wittgenstein, only to introduce Wittgenstein immediately afterwards which forced upon him the, by far, greatest reorganization of *Proto-Structure*: to split one chapter on normal science into two. In addition, Kuhn says in a taped interview, although much later (1995) than in D.G. Cedarbaum's reported conversation of 26 November 1979 (Cedarbaum 1983, p. 188 fn. 83), that initially he was not aware of the parallel of his paradigms to Wittgenstein (Kuhn et al. [1997] 2000, p. 299). It appears most likely that it was Stanley Cavell who made Kuhn aware of the usefulness of Wittgenstein for the analysis of normal science. Kuhn had sent or given *Proto-Structure* to Cavell and he acknowledges in the Preface to *Structure* that he was able "to point me the way through or around several major barriers encountered while preparing my first manuscript" (Kuhn [1962] 1970, p. xi). Among the people who Kuhn mentions in the Preface, Cavell is the one most intimately engaged with the late Wittgenstein in the early sixties. However, it is only a conjecture that Cavell made Kuhn aware of Wittgenstein at this critical point. But it is interesting enough that Wittgenstein plays no significant role in *Proto-Structure*.

---

[4] This phrase is in *Proto-Structure* on p. 14; it is also in *Structure* (Kuhn (1970 [1962])) on p. 14 and in Kuhn (1964), written in 1961, on p. 354. The absence of Wittgenstein in *Proto-Structure* has recently also been noted by another author: Isaac (2012), pp. 105–106. However, it is not entirely clear that what Isaac calls the "penultimate draft" of *Structure*, supposedly finished in summer 1960, is identical with *Proto-Structure* which was finished in the spring of 1961. In an unpublished working draft, Matteo Collodel convincingly argues that there was an even earlier draft of *Structure* than *Proto-Structure*. Collodel calls this earlier draft *Proto-Proto-Structure* and dates its completion at September 1960: see Collodel (2013, p. 6 fn. 17).

[5] This story goes back to Cedarbaum (1983, p. 188).

13.2.2   *Context of Discovery vs. Context of Justification*

In the introductory chapter of *Structure*, there is a well-known passage regarding the distinction between the context of discovery and the context of justification. This passage is well-known because it is a vital element in the critical discussion of this distinction that emerged in the early 1960s.[6] Before that time, the distinction played a constitutive role for standard philosophy of science: it clearly defined the area for which philosophy of science was legitimized: the context of justification. The context of discovery was the province of the empirical disciplines like history or sociology or psychology of science. However, Kuhn does not seem to respect the fundamental divide between the two contexts, and he is obviously completely aware of that:

> In the preceding paragraph, I may even seem to have violated the very influential contemporary distinction between the "context of discovery" and "the context of justification." Can anything more than profound confusion be indicated by this admixture of diverse fields and concerns?
> Having been weaned intellectually on these distinctions and others like them, I could scarcely be more aware of their import and force. (Kuhn [1962] 1970, pp. 8–9)

Kuhn defends himself against anticipated criticism of his apparent disrespect for the context distinction by one of the most enigmatic sentences of *Structure*:

> Rather than being elementary logical or methodological distinctions, which would thus be prior to the analysis of scientific knowledge, they now seem integral parts of a traditional set of substantive answers to the very questions upon which they have been deployed. (Kuhn [1962] 1970, pp. 8–9)[7]

Again, in order to understand this convoluted sentence, one might turn to *Proto-Structure* seeking help. However, in *Proto-Structure* there is nothing at all about the context distinction. The reference to the context distinction is therefore a late addition to *Structure*, after April 1961. Puzzled about his remark about the context distinction in *Structure*, I asked Kuhn about it. He told me in 1984 that the passage on the context distinction was a "throw-away remark"—Cavell had made him aware that people might object to his book by invoking the context distinction.

13.2.3   *Intermediate Summary on Proto-Structure*

Kuhn's development from *Proto-Structure* of 1961 to *Structure* of 1962 comprises two important and novel ingredients: Kuhn's introduction of the late Wittgenstein's theory of concepts and Kuhn's reaction to anticipated criticisms regarding the context distinction. When reading *Structure*, most readers probably assumed that these two elements of *Structure* were constitutive ingredients, governing it more or less from its inception. However, in truth they were late additions to *Proto-Structure*, not at all

---

[6] See, e.g., Hoyningen-Huene (1987).
[7] I have dealt with this sentence in detail in Hoyningen-Huene (2006a, pp. 124–126).

inspiring the overall composition of *Structure*, and both very probably triggered by Cavell.

13.3 The Late Kuhn

At least from the early 1980s on, Kuhn worked on a book in which, among other things, he tried to base incommensurability on new foundations. During this time, he saw himself involved in a rather rapid and profound development. When I arrived at Boston in August 1984, Kuhn was, in principle, pleased with my book project that set out to reconstruct his philosophy including its development.[8] However, as he told me later, he thought that my project was somewhat unfortunate insofar as it seemed to him ill-timed. As I had planned to finish my project during the second half of the 1980s, I could not possibly cover the work that Kuhn was currently involved. He hoped at the time of our first encounter that he could finish his new book at least by the late 1980s.

13.3.1 The Plurality of Worlds

Kuhn's book project had different titles at different times: in December 1984, it was entitled *Scientific Development and Lexical Change*; in March 1990 it was called *Words and Worlds: An Evolutionary View of Scientific Development* and the final title was *The Plurality of Worlds: An Evolutionary Theory of Scientific Development*. I may say that I liked the last title best because it picks up what I called the plurality-of-phenomenological-worlds thesis", which I assessed as a "fundamental assumption of Kuhn's theory" (Hoyningen-Huene 1993, p. 36).

The Table of Contents of *The Plurality of Worlds*, in the version of 1994, had significantly changed in comparison to the version of 1990 and consisted of three parts, each comprising three chapters:

Preface
- Part I: The Problem
    1. Scientific Knowledge as Historical Product
    2. Breaking into the Past
    3. Taxonomy and Incommensurability
- Part II: A World of Kinds
    4. Biological Prerequisites to Linguistic Descriptions: Tracks and Situations
    5. How Kind-Terms Mean
    6. Singletons and the Entry of Scientific Laws
- Part III: Reconstructing the World
    7. Looking Backward and Moving Forward
    8. Theory-Choice and the Nature of Progress
    9. What's in a Real World?
Epilogue

---

[8] See Kuhn's Foreword to Hoyningen-Huene (1993, pp. xi–xiii).

Unfortunately, only Chaps. 2 through 6 exist (in manuscript form). Of Chaps. 2–4, I own a version from September 19, 1994. Of Chap. 5, my version is dated October 11, 1994; my version of Chap. 6 is, according to my notes, from August 1995. I have discussed these existing chapters in several sessions consisting each of several hours during my last stay with Kuhn between September 2 and 11, 1995 in Kuhn's home in Cambridge, MA. The book manuscript is not publicly available. Shortly before his death in June 1996, Kuhn had asked two younger colleagues to edit the book manuscript and complement the missing chapters on the basis of his notes and his oral suggestions. One of the potential editors, John Haugeland, died in 2010. The other potential editor is James Conant. In the bygone 17 years until today (Sept 2013), Conant's homepage has been featuring Kuhn's *The Plurality of Worlds* as "forthcoming".[9]

Kuhn treated some of the topics of *The Plurality of Worlds* also in publications from the 1980s and 1990s, most centrally in Kuhn (1991) and Kuhn (1992), but also in Kuhn (1981), Kuhn (1983), Kuhn (1989), Kuhn (1990), and Kuhn (1993). In addition to these easily accessible sources, there are manuscripts of three lecture series that Kuhn had given and that circulated fairly widely. In November 1980, Kuhn gave the Perspective Lectures at Notre Dame entitled "The Nature of Conceptual Change"; in November 1984, he gave the Thalheimer Lectures at Johns Hopkins entitled "Scientific Development and Lexical Change", and in November 1987, he gave the Shearman Memorial Lectures at University College London entitled "The Presence of Past Science". In addition, in 1990 he gave a pertinent talk to the Cognitive Science Colloquium at UCLA entitled "An Historian's Theory of Meaning".[10]

13.3.2 The Train of Thought of the Plurality of Worlds

Here is a very condensed presentation of the book's train of thought. Kuhn begins by noting the evident fact that narratives in the history of science need a starting point. This starting point typically concerns the state of the art of a specific scientific area with which the narrative is to deal. As the historical starting point is described in the relevant historical sources of the time in their language and is addressed to their contemporaries, our reading and understanding of these sources requires interpretation. This interpretation typically recovers a set of interrelated kind concepts characteristic of that science, which differs in structure from the corresponding set of contemporary kind concepts. "Differing in structure" means that this older set of kind concepts implies a taxonomy of things, processes, etc. that differs from the taxonomy we are used to today. This difference of taxonomies is incommensurability. In order to better understand taxonomies and their structural differences, i.e.

---

[9] http://philosophy.uchicago.edu/faculty/conant.html, last accessed Sep 6, 2013. I may add that over the years I asked Conant via email again and again when the book would be published but I was always promised jam tomorrow.

[10] Some of the information in this paragraph can be found in Marcum (2005, pp. 23–26).

incommensurability, a theory of kind concepts is necessary. This ends part I and at leads over to part II.

Kind terms have, according to Kuhn, a biological foundation. It is obvious that many animals show behavioral differences with regard to things belonging to different kinds, for example food vs. non-food, con-specific vs. member of a different species, etc. The biological foundation of kind terms comes to the fore in humans in their very early ability to recognize individual people. What has this ability to do with kind terms? To recognize someone means to re-identify him or her, and that means to comprehend the different appearances of the respective person as appearances of a single person. In other words, all the different appearances belong to the same kind. Thus, individual recognition and the ability to classify things as belonging to the same kind are close relatives.

Taxonomic kind terms come in different independent hierarchies in which terms further down exhibit lesser generality than the terms above them. The most important hierarchies concern things and materials. The taxonomic kind terms in any given hierarchy are not introduced by definitions, but by similarities between entities belonging to the same kind, and differences to the members of another kind. The most important principle governing any hierarchy of taxonomic kind terms is the no overlap principle. It states that the extensions of terms within any given hierarchy must not overlap, i.e., they must either have an empty intersection or one of the extensions is a subset of the other. Clearly, this principle derives from the aim of any taxonomy, namely, to assign any given entity that belongs to the hierarchy to exactly one kind at the lowest level, and to assign any kind to only one kind at the next higher level. It is very important to note that the no overlap principle only holds within any given hierarchy of kind terms; it does not hold between kind terms belonging to different hierarchies. The *structure* of a hierarchy of kind terms is the totality of the relationships among the extensions of the terms in the hierarchy; it is also called the structure of the respective *lexicon* (of kind terms). Members of a language community share the structure of the lexicon of their kind terms. However, different members of that community may use widely different criteria for how to identify the members of any given kind. These differences are usually hidden in normal conversation as long as the structure of the lexicon is preserved. What Kuhn used to call "normal science" is now a phase of scientific development in which the structure of the pertinent lexicon of kind terms is stable. Correspondingly, revolutionary developments are those in which the integrity of the structure of a part of some lexicon is threatened. The result of a successful revolutionary development is a lexicon whose structure is somewhat modified in comparison to the old lexicon; in addition, some new kind terms may have been introduced and some old kind terms abandoned. Note that these changes are always *local*: they concern a certain part of the lexicon, usually only a few interrelated terms.

Kuhn made huge efforts to establish this theory of kind terms because it laid the foundations for a thorough understanding of incommensurability, i.e., differences in the structure of lexicons of taxonomic kind terms. As he had already indicated in *Structure*, incommensurability understood in such a way had fundamental implications for our understanding of the progress of science: scientific progress could not possibly be an approach to truth.[11] Furthermore, Kuhn was always troubled by the idea that a purely *instrumentalist* understanding of scientific development seemed to eliminate the idea of a serious competition between different theories (or models) due to mutual contradiction. An instrumentalist may use any theory or model she pleases depending on context; whether the models contradict each other in cases of overlapping applications is irrelevant as long as their predictive power is unscathed. For the realist, however, theories about the same domain stand in competition about (approximate) truth. The winning theory in such competitions then represents the best suited candidate for approximate truth. Because the (natural) sciences have so far been very successful in almost unanimously identifying the winning theories, these sciences gained an almost totally unrivalled cognitive authority. However, this cognitive authority seems to depend on a realist understanding of science; an instrumentalist (or social constructivist) understanding of science seems to unavoidably undermine this cognitive authority. It was one of Kuhn's fundamental goals to reconcile the idea of a justified cognitive authority of the sciences with a denial of their realist interpretation. Incommensurability, properly understood, would show how these seemingly irreconcilable ideas are indeed two sides of one coin. Furthermore, such an understanding of incommensurability would lead to a deeply transformed idea of what reality is. Unfortunately, part III of *The Plurality of Worlds* in which these questions would be treated is unwritten.

13.3.3 What is New?

Let me now quickly highlight the element of *The Plurality of Worlds* that seems to be most novel, relative to what Kuhn had published earlier. Clearly, this is Kuhn's (sketch of) a theory of kind terms. It is a thoroughly naturalistic theory, strongly multidisciplinary, drawing heavily on developmental psychology. Its core element, structured kind sets, is designed to play the role that paradigms played, however imperfectly, in *Structure*. However, despite its naturalistic origin, this theory of kind terms is supposed to found a quasi-transcendental theory of experience and of reality.[12] It is the structure of the lexicon that determines the range of possible experiences and, at the same time, what we (legitimately) take as real. The foundational role of the structure of the lexicon for the range of possible experiences and for reality does, however, *not* preclude that the lexicon can be changed due to particular experiences; the apriori is therefore not absolute, but relativized. Significant anomalies violating the no overlap principle of one of the hierarchies may force a (local) change in a hitherto unchallenged lexical structure.

---

[11] For further discussion on Kuhn's criticism of truth in science, see Devlin, Chap. 11, in this volume.

[12] I have also tried to develop this point in Hoyningen-Huene (2008).

13.4 Conclusion

Kuhn was intellectually never at rest although he had discovered the central topic
of his philosophical thinking very early in his career, in 1947. By a sudden change
in his perception of a text by Aristotle, he caught a glimpse of incommensurability.[13] This remained the central topic of his life. During a very short period from
1961 to 1962, he added the Wittgensteinian twist to his idea of incommensurability.
During the long period from the early 1980s to his death in 1996, he added naturalistic underpinnings by drawing on biology and developmental psychology. The full
philosophical consequences of the latter move remain to be drawn.

References

Cedarbaum, D. G. 1983. Paradigms. *Studies in History and Philosophy of Science* 14:173–213.
Collodel, M. 2013. *A note on Kuhn and Feyerabend at Berkeley (1956–1964)*. Working draft, July
    31, 2013.
Hoyningen-Huene, P. 1987. Context of discovery and context of justification. *Studies in History
    and Philosophy of Science* 18:501–515.
Hoyningen-Huene, P. 1993. *Reconstructing scientific revolutions: Thomas S. Kuhn's philosophy of
    science*. Chicago: University of Chicago Press.
Hoyningen-Huene, P. 2006a. Context of discovery versus context of justification and Thomas
    Kuhn. In *Revisiting discovery and justification: Historical and philosophical perspectives on
    the context distinction,* ed. J. Schickore and F. Steinle, 119–131. Dordrecht: Springer.
Hoyningen-Huene, P. 2006b. More letters by Paul Feyerabend to Thomas S. Kuhn on *proto-
    structure*. *Studies in History and Philosophy of Science* 37:610–632.
Hoyningen-Huene, P. 2008. Commentary on Bird's paper. In *Rethinking scientific change and
    theory comparison: Stabilities, ruptures, incommensurabilities?* ed. L. Soler, H. Sankey, and P.
    Hoyningen-Huene, 41–46. Dordrecht: Springer.
Isaac, J. 2012. Kuhn's education: Wittgenstein, pedagogy, and the road to structure. *Modern
    Intellectual History* 9 (1): 89–107.
Kuhn, T. S. [1962]1970. *The structure of scientific revolutions*. 2nd ed. Chicago: University of
    Chicago Press.
Kuhn, T.S. 1964. A function for thought experiments. In *L'aventure de la science. Mélanges
    Alexandre Koyré. Vol. 2.,* 307–334. Paris: Herman (reprinted in *The Essential Tension*, Chicago:
    University of Chicago Press, 1977).
Kuhn, T. S. 1981. *What are scientific revolutions? Occasional Paper #18*. Cambridge: Center for
    Cognitive Science, M.I.T. (Reprinted in L. Krüger, L. J. Daston, and M. Heidelberger, eds., *The
    probabilistic revolution, vol. 1: Ideas in history,* 7–22. Cambridge: M.I.T. Press. 1987).
Kuhn, T. S. 1983. Commensurability, comparability, communicability. In *PSA 1982: Proceedings
    of the 1982 Biennial Meeting of the Philosophy of Science Association, vol. 2.,* ed. P. D. Asquith
    and T. Nickles, 669–688. East Lansing: Philosophy of Science Association.
Kuhn, T. S. 1989. Possible worlds in history of science. In *Possible worlds humanities, arts, and
    sciences. Proceedings of Nobel Symposium 65,* ed. S. Allén, 9–32. Berlin: de Gruyter.
Kuhn, T. S. 1990. Dubbing and redubbing: The vulnerability of rigid designation. In *Scientific
    theories. Minnesota studies in the philosophy of science Vol. XIV.,* ed. C. W. Savage, 298–318.
    Minneapolis: University of Minnesota Press.

---

[13] See, e.g., Kuhn (1981, pp. 8–12).

Kuhn, T. S. 1991. The road since structure. In *PSA 1990, Proceedings of the 1990 Biennial Meeting of the Philosophy of Science Association*, ed. A. Fine, M. Forbes, and L. Wessels, 3–13. East Lansing: Philosophy of Science Association.

Kuhn, T. S. 1992. *The trouble with the historical philosophy of science. Robert and Maurine Rothschild Distinguished Lecture, 19 November 1991*. Cambridge: An Occasional Publication of the Department of the History of Science, Harvard University. (Reprinted in Kuhn, T. S. 2000. *The road since structure. Philosophical essays, 1970–1993, with an Autobiographical Interview*, ed. J. Conant and J. Haugeland. Chicago: University of Chicago Press).

Kuhn, T. S. 1993. Afterwords. In *World changes: Thomas Kuhn and the nature of science*, ed. P. Horwich, 311–341. Cambridge Mass: MIT Press.

Kuhn, T. S., A. Baltas, K. Gavroglu, and V. Kindi. 1997. A discussion with Thomas S. Kuhn: A Physicist who became a historian for philosophical purposes. Neusis 6 (Spring-Summer):145–200. (Reprinted in Kuhn, T. S. 2000. *The road since structure: Philosophical essays, 1970–1993, with an autobiographical interview*, ed. J. Conant and J. Haugeland, 253–323. Chicago: University of Chicago Press.

Marcum, J. A. 2005. *Thomas Kuhn's revolution: An historical philosophy of science*. London: Continuum.

Index

"The Function of Measurement in the Modern
    Physical Sciences', 47
Criticism and the Growth of Knowledge, 18
E. coli., 73, 74
Leviathan and the Air-Pump, 30, 92, 98
Little Science, Big Science, 17
Structure of Scientific Revolutions, 11, 32,
    91, 135
The Essential Tension, 41
The Road Since Structure, 135, 140

A
a priori, 141, 163
Abnormal science, 99
Absolute spirit, 27
Anomaly, 27, 28, 158
Archimedean platform, 157, 159
Aristotle, 138, 143
Axioms of probability, 72

B
Bacon, F., 34
Barnes, B., 18, 19
Bayesian, 71
Beiser, F., 25
Belief, 28, 29
Biology, 35
Bird, A., 27
Bloor, D., 18
Bridge principles, 43

C
Cassirer, E., 49
Cavell, S., 12
Changing Order, 92
Classical mechanics, 150

Collins, Harry, 18, 92, 94
Comte, A., 25
Conditions of possibility, 13, 15, 19, 141,
    146, 147, 149, 150
Conferences, 98
Confirmation, 26, 27
Conservative, 25–28, 36
Copernican revolution, 30
Crisis, 13, 28, 30, 31, 71, 158
Cross-induction, 75, 79, 86, 89

D
Darwin, C., 35, 46
Daston, L., 40, 95
Demarcation, 14, 93
Descartes, R., 23, 138
Determinism, 34, 35, 37
Determinist, 25, 27, 28, 35
Different worlds, 142, 145, 159
Disciplinary matrix, 26, 158
Discovery/justification distinction, 43, 45
Double goal, 154, 155, 162

E
Edinburgh school, 29, 91
Empirical validation, 47
Engineers, 96–101
Epistemic access, 159, 164
Evolution of science, 30
Exemplar, 26, 158
Externalism, 23, 31, 32, 35, 36
Extraordinary science, 32, 33

F
Falsification, 36, 145
Falsificationism, 36
Feigl, H., 47

© Springer International Publishing Switzerland 2015
W. J. Devlin, A. Bokulich (eds.), *Kuhn's Structure of Scientific Revolutions—50 Years On*,
Boston Studies in the Philosophy and History of Science 311,
DOI 10.1007/978-3-319-13383-6

Feyerabend, P., 46
Fleck, L., 14, 100
Forces and Fields, 42
Forman, P., 15, 18
Friedman, M., 40

G
Galileo, G., 138, 142, 143
Galison, P., 14, 15, 82, 83, 97
Gavagai, 146
Generalizations, 44, 48, 71, 73, 74, 158
Gestalt, 12, 138, 143, 148
Giere, R., 40, 139–141, 144, 151
Gillispie, C., 11
Gravitational radiation, 92, 94

H
Hacking, I., 145
Hanson, N.R., 137
Hegel, G.W.F., 23, 34, 35
Heilbron, J.L., 18
Herder, J.G., 24
Historical facts, 43, 72
Historicism, 23–25, 27–29, 34–37
Historiography, 27, 29
History of science, 11, 14, 18, 19, 23, 26, 29,
        30, 35–37, 43, 49, 72, 89, 154
Holism, 44
Hollinger, D., 14, 72
Hoyningen-Huene, P., 135, 136, 138, 144
Hypothesis, 81, 83, 85

I
Induction, 48, 72, 74, 76, 78, 80, 87
        meta-, 77, 154
Inductivist, 36, 74
Inference, 74, 77, 78, 86, 146
Internalism, 30, 32, 35
Irrationalism, 37

K
Kaiser, D., 15, 97
Kant, I., 24, 49, 136
Kantian
        neo-, 49, 148
Kepler, 28, 31, 35
Kevles, D., 15
Kind terms, 140, 145
Kind-changes, 149, 151
Knowledge
        scientific, 34, 44, 49, 97, 154, 160
        tacit, 12, 14, 92, 97
        technical, 96, 97
Knowledge of nature, 154, 161, 165

L
Lab scientists, 98
Lab studies, 93
Laboratory ethnography, 93
Lakatos, I., 13
Language
        functionalist, 49
Leibniz, G.W., 139, 149
Lexicon, 142, 143
Logic, 12, 26, 27, 37, 88
Logical empiricism, 27, 36, 43
Luminiferous ether, 72

M
Mannheim, K., 25
Marx, K., 25
Marxism, 15
Massimi, M., 136, 149
Merton, R., 14, 16, 30
Metaphor, 40, 43, 83, 164
Metaphysics, 136, 164
Meyerson, E., 49
Mind-dependent world, 163–165
Mind-independent world, 153, 156, 164, 166
Mody, C., 95, 97–99

N
Narrative, Romantic *See* Romantic
        literature, 49
National Science Foundation (NSF), 15
Natural kinds, 140, 145, 147, 148
Naturalism, 13, 15, 18, 19, 138, 148
        challenges of, 144
Naturalized Kantian kinds (NKKs), 136, 144,
        146, 147, 151
Neutral meta-language *See* Language, 47
Neutral-arbiter problem, 82
Newton, I., 28, 138
Newtonian mechanics, 140, 141, 150
Newtonian tradition, 26
Nietzsche, N., 139
Nominalism, 145
Normal science, 13, 28, 29, 32, 36, 43, 91, 93,
        94, 98, 153, 158
Normative tasks, 45

O
Objective idealism, 24
Objectivism, 34
Objectivity, 12, 95

P
Paradigm, 12, 17, 26–28, 31–33, 73, 82, 96,
        136, 158, 160
Paradigm change, 92

Paradigm shift, 82, 83, 137, 160
Perception, 17
Pessimistic induction (PI), 72, 79, 81, 87
Phenomena
    challenge of, 136
Philosophy of science, 11, 36, 40–42,
    47, 154
Phlogiston, 72, 81, 145
Physics, 76, 87, 100
Polanyi, M., 12, 14
Popper, K., 14
Postscript, 18, 153
Pragmatism, 144
President Eisenhower's Farewell
    Address, 16
Price, D., 17
Problem of inconsistency, 154, 155, 161,
    163, 165
Psychology, 12, 137, 148
Ptolemaic system, 96
Putnam, H., 139, 157
Puzzle solving, 28, 41, 158

Q
Quantum mechanics, 76

R
Rationality, 18, 24, 37, 46, 73, 82, 89
    principles of, 71
Rationality of science See Scientific
    rationality, 37
Realism
    mild, 135, 136
    perspectival, 139, 140, 144
Reality See Realism, 11
Reconstruction, 44
Reductionism, 93
Reichenbach, H., 14, 75, 163
Relativism, 18, 24, 34, 136, 150
    alethic, 144, 148, 150
    conceptual, 136, 142–144, 149
Reliability, 78–82, 85, 86, 90
Representational theory of mind, 48
Research communities, 95
Revolutionary science, 27, 31, 32, 36, 91
Romantic literature, 40, 49
Roush, S., 78, 84, 85
Rules, 12, 27, 71, 79

S
Scientific change See Paradigm
    change, 27

Scientific community, 14, 15, 28
Scientific method, 16, 19, 36, 79, 87
Scientific perspectivism, 139
Scientific progress, 12, 18, 19, 41
Scientific rationality, 12, 37, 139
Scientific revolution, 28, 33, 137, 147
Scientific theory, 26, 43
Shapin, S., 14, 15, 30, 100
Skepticism, 83, 92, 164
Social constructivism, 29
Sociology, 19
Sociology of science, 29
Sociology of Scientific Knowledge (SSK), 23,
    29, 91
Standard Model, 87–89
Stanford, K., 83
Strong Programme, 19, 29, 30
Sub-normal science, 99, 100

T
Taxonomy, 142
Technology, 14, 30, 34, 41, 50, 100, 101
The Steelman Report, 15
Theory-dependence of observation, 159
Theory-laden See Theory-dependence of
    observation, 82
Theory-ladenness of observation, 154
Truth
    correspondence theory of, 153, 155,
        163, 165
    perspectival, 136
    phenomenal-world correspondence theory
        of, 163, 164, 166

U
Unity of science, 12, 14
Universalists
    transhistorical, 46
Unobservables, 72, 83, 85, 86
    claims, 85
    entities, 72

V
Values, 24, 29, 30, 42, 158
Vico, G., 24

W
Weinberg, A., 16
Wittgenstein, L., 12
World-changes, 136, 138, 145, 149, 151

Printed in the United States
By Bookmasters